ISBN 978-1-330-30035-0
PIBN 10018859

1 MONTH OF
FREE
READING

at

www.ForgottenBooks.com

By purchasing this book you are eligible for one month membership to ForgottenBooks.com, giving you unlimited access to our entire collection of over 1,000,000 titles via our web site and mobile apps.

To claim your free month visit: www.forgottenbooks.com/free18859

Cambridge Natural Science Manuals.

GEOLOGICAL SERIES.

THE PRINCIPLES

OF

STRATIGRAPHICAL GEOLOGY

London: C. J. CLAY AND SONS,
CAMBRIDGE UNIVERSITY PRESS WAREHOUSE,
AVE MARIA LANE.

AND

H. K. LEWIS,
136, GOWER STREET, W.C.

Leipzig: F. A. BROCKHAUS.
New York: THE MACMILLAN COMPANY.
Bombay: E. SEYMOUR HALE.

THE PRINCIPLES

OF

STRATIGRAPHICAL GEOLOGY

BY

J. E. MARR, M.A., F.R.S.

FELLOW AND LECTURER OF S. JOHN'S COLLEGE, CAMBRIDGE,
AND UNIVERSITY LECTURER IN GEOLOGY.

CAMBRIDGE:
AT THE UNIVERSITY PRESS.
1898

𝔊𝔞𝔪𝔟𝔯𝔦𝔡𝔤𝔢:

PRINTED BY J. & C. F. CLAY,

AT THE UNIVERSITY PRESS.

PREFACE.

THE present work has been written in order that students may gain by its perusal some idea of the methods and scope of Stratigraphical Geology. I believe that this idea can be obtained most satisfactorily, if a large number of the details connected with the study of the stratified rocks are omitted, and I have accordingly given very brief accounts of the strata of the different Systems.

The work is intended for use in conjunction with any book which treats of the strata of the Geological Column at considerable length; some of these books are mentioned on pages 124, 125.

J. E. M.

CAMBRIDGE,
    November, **1898.**

CONTENTS.

CONTENTS.

CHAPTER XV.

ADDENDA ET CORRIGENDA.

p. 38, line 15 from bottom: for 'joining' read 'jointing'.

p. 208, line 6 from bottom: for 'Dr' read 'Messrs Medlicott and

p. 214, line 15 from bottom: after 'Fermo-Carboniferous Strata' insert 'through the Permian'

p. 217, last line of footnote: for 'Dr' read 'Messrs Medlicott and'

,, insert a second footnote: 'For information concerning the Permian volcanic rocks see Sir A. Geikie's *Ancient Volcanoes of Great Britain.*'

p. 235, insert a footnote: 'A good account of the British Jurassic rocks will be found in Mr H. B. Woodward's Memoir on "The Jurassic Rocks of Britain." *Mem. Geol. Survey,* 1893—.'

p. 250, top line: for 'Gardiner' read 'Gardner'

CHAPTER I.

INTRODUCTION.

It is the aim of the Stratigraphical Geologist to record the events which have occurred during the existence of the earth in the order in which they have taken place. He tries to restore the physical geography of each period of the past, and in this way to write a connected history of the earth. His methods are in a general way similar to those of the ethnologist, the archæologist, and the historian, and he is confronted with difficulties resembling those which attend the researches of the students of human history. Foremost amongst these difficulties is that due to the imperfection of the geological record, but similar difficulty is felt by those who pursue the study of other uncertain sciences, and whilst this imperfection is very patent to the geologist, it is perhaps unduly exaggerated by those who have only a general knowledge of the principles and aims of geology.

The history of the earth, like other histories, is a connected one, in which one period is linked on to the next. This was not always supposed to be the case; the catastrophic geologist of bygone times believed that after each great geological period a convulsion of nature left the earth's crust as a *tabula rasa* on which a new set of records was engraved, having no connexion with those which had been destroyed. Careful study of the records of the rocks has proved that the conclusions of the catastrophists were erroneous, and that the events of one period produce their impression upon the history of the next. Every event which occurs, however insignificant, introduces a new complication into the conditions of the earth, and accordingly those conditions are never quite the same. Although the changes were no doubt very slow, so that the same general conditions may be traced as existent during two successive periods, minor complications occurred in the inorganic and organic worlds, and we never get an exact recurrence of events. Vegetable deposits may now be in process of accumulation which in ages to come may be converted into coal, but the general conditions which were prevalent during that Carboniferous period when most of our workable coal was deposited do not now exist, and will never exist again. The changes which have taken place and which are taking place show an advance from the simple to the more complex, and the stratigraphical geologist is confronted with a problem to which the key is development, and it is his task to trace the development of the earth from the primitive state to the complex condition in which we find it at the present day.

Our general ignorance of the events of the earliest periods of the history of the earth will be emphasised in the sequel, and it will be found that the complexity which marks the inorganic and organic conditions which existed during the deposition of the earliest rocks of which we have detailed knowledge points to the lapse of enormous periods of time subsequent to the formation of the earth, and previous to the deposition of those rocks. The imperfection of the record is most pronounced for that long period of time, but in this respect the geologist is in the same condition as the student of human history, for the relics of the early stone age prove that man in that age had attained a fairly high state of civilisation, and the gap which separates palæolithic man from the first of our race is relatively speaking as great as that which divides the Cambrian period from the commencement of earth-history. Nevertheless, human history is a science which has made gigantic strides towards the solution of many problems connected with the development of man and civilisation, and similarly geology has advanced some way in its task of elucidating the history of our globe.

The task of the stratigraphical geologist is two-fold. In the first place, he must establish the order of succession of the strata, for a correct chronology is of paramount importance to the student of earth-lore. The precautions which must be taken in making out the order of deposition of the rocks of any area, and correlating those of one area with those of another will be considered in the body of the work. When this task is completed, there yet remains the careful examination of all the information supplied by a study of the rocks of the crust, in order to ascertain the actual conditions which existed during the deposition of any stratum or group of strata. In practice, it is generally very difficult to separate these two departments of the labour of the stratigraphical geologist, and the two kinds of work are often done to a large extent simultaneously, or sometimes alternately. Frequently the general succession of the deposits comprising an important group is ascertained, and at the same time observations made concerning the physical characters of the deposits and the nature of their included organisms, which are sufficient to afford some insight into the general history of the period when these deposits were laid down; a more detailed classification of the same set of deposits may be subsequently made, and as the result of this, more minute observations as to the variations in the physical and biological conditions of the period are possible, which permit us to write a much more concise history of the period. So great has been the tendency to carry on work in a more and more detailed manner, that it is very difficult if not impossible to tell when any approach to finality is reached in the study of a group of strata in any area. Roughly speaking, we may state that our knowledge of a group of strata is obtained by three processes, or rather modifications of one process. The general order of succession is established by the pioneer, frequently as the result of work carried on through one or two seasons. Subsequently to this, a more minute subdivision of the rocks is possible as the result of labours conducted by one or more workers who are enabled to avail themselves of the work of the pioneer, and our knowledge of the rocks is largely increased thereby. But the minutiæ, often of prime importance, are supplied by workers who must spend a large portion of their time in the area where the work lies, and it is only in districts where work of this character has been performed, that our knowledge of the strata approaches completion. The strata of the Arctic regions, for example, have in many places been examined by pioneers, but a great deal remains to be done in those regions; the main subdivisions only have been defined in many cases, and our information concerning the physical history of Arctic regions in past times is comparatively meagre. To come nearer home—a few miles north of Cambridge lies the little patch of Corallian rock at

Upware; it has been frequently visited, and a large suite of organic remains extracted from it, but no one has devoted the time to the collection of remains from this deposit which has been devoted to that of some other formations presently to be mentioned, and accordingly our knowledge of the fauna of that deposit is far from complete. Contrast with this the information we possess of the little seam known as the Cambridge Greensand, from which organic remains have been sedulously collected during the extensive operations which have been carried on for the extraction of the phosphatic nodules which occur in the seam. The suite of relics of the organisms of that period is accordingly far more perfect than in the case of many other beds, and indeed the large and varied collection of relics of the vertebrata of the period which furnish much information of value to the palæontologist would not have been gathered together, had not this seam been so carefully worked, and an important paragraph in the chapter bearing on the history of this period would have remained unknown to us. Again, two little patches of limestone of the same age, one in central England and the other in the island of Gothland, have been the objects of sedulous inquiry by local observers, and we find again that our knowledge of the physical history of the period, as regards these two regions, is exceptionally perfect. Special stress is laid upon this point, for in these days, when every county possesses its learned societies whose members are desirous of advancing in every possible way the progress of science, it is well to insist upon the importance of this detailed work which can only be done by those who have a large amount of time to devote to the rigorous examination of the rocks of a limited area.

CHAPTER II.

ACCOUNT OF THE GROWTH AND PROGRESS OF STRATIGRAPHICAL GEOLOGY.

THE history of the growth of a science is not always treated as an essential part of our knowledge of that science, and many text-books barely allude to the past progress of the science with which they deal. The importance of a review of past progress has, however, attracted the attention of many geologists, and Sir Charles Lyell, in his *Principles of Geology,* gave prominence to an historical sketch of the rise and progress of the science. Historical studies of this nature have more than an academic value; the very errors made by men in past times are useful as warnings to prevent those of the present day from going astray; the lines along which a science has progressed in the past may often be used as guides to indicate how work is to be conducted in the future; but perhaps the greatest lesson which is taught by a careful consideration of the rise and progress of a study is one which has a moral value, for he who pays attention to the growth of his science in past times, gains a reverence for the old masters, and at the same time learns that a slavish regard for authority is a dangerous thing. This is a lesson which is of the utmost importance to the student who wishes to advance his science, and will prevent him from paying too little attention to the work of those who ·have gone before him, whilst it will enable him to perceive that as great men have fallen into error through not having sufficient data at their disposal, he need not be unduly troubled should he find that conclusions which he has lawfully attained after consideration of evidence unknown to his predecessors clash with those which they adopted. Want of this historic knowledge has no doubt caused many workers to waste their time on work which has already been performed, but it has also led others to withhold important conclusions from their fellow-workers because they were supposed to be heterodox. In an uncertain science like geology one of the great difficulties is to keep an even balance between contempt and undue respect for authority, and assuredly a scientific study of the past history of a science will do much to enable a student to attain this end. It will be useful, therefore, at this point to give a brief account of the rise and progress of the study of stratigraphical geology, so far as that can be done without entering into technical details, at the same time recommending the student to survey the progress of this branch of our science for himself, after he has mastered the principles of the subject, and such details as are the property of all who have studied the science from the various text-books written for advanced students.

William Smith, the 'Father of English Geology,' is rightly regarded as the founder of stratigraphical geology on a true scientific basis, but like all great discoverers, his work was foreshadowed by others, though so dimly, that this does not and cannot detract from his fame. It is desirable, however, to begin our historical review at a time somewhat further back than that at which Smith gave to the world his epoch-making map and memoirs.

Before the eighteenth century, stratigraphical geology cannot be said to have existed as a branch of science—the way had not been prepared for it. Data had been accumulated which would have been invaluable if at the disposal of open-minded philosophers, but with few exceptions prejudice prevented the truth from becoming known. There were two great stumbling-blocks to the establishment of a definite system of stratigraphical geology by the writers of the Middle Ages, firstly, the contention that fossils were not the relics of organisms, and, secondly, when it was conceded that they represented portions of organisms which had once existed, the assertion that they had reached their present positions out of reach of the sea during the Noachian Deluge. For full details concerning the mischievous effects of these tenets upon the science the reader is referred to the luminous sketch of the growth of geology in the first four chapters of Sir Charles Lyell's *Principles of Geology.*

The disposition of rocks in strata, and the occurrence of different fossils in different strata, was known to Woodward when he published his *Essay toward a Natural History of the Earth* in 1695, and the valuable collections made by Woodward and now deposited in the Woodwardian Museum at Cambridge, show how fully he appreciated the importance of these facts, though he formed very erroneous conclusions from them, owing to the manner in which he drew upon his imagination when facts failed him, maintaining that fossils were deposited in the strata according to their gravity, the heaviest sinking first, and the lightest last, during the time of the universal deluge. The following extracts from Part II. of Woodward's book, show the position in which our knowledge of the strata stood at the end of the seventeenth century: "The Matter, subsiding..., formed the *Strata* of Stone, of Marble, of Cole, of Earth, and the rest; of which Strata, lying one upon another, the Terrestrial Globe, or at least as much of it as is ever displayed to view, doth mainly consist.... The Shells of those Cockles, Escalops, Perewinkles, and the rest, which have a greater degree of Gravity, were enclosed and lodged in the *Strata* of Stone, Marble, and the heavier kinds of Terrestrial Matter: the lighter Shells not sinking down till afterwards, and so falling amongst the lighter Matter, such as Chalk, and the like...accordingly we now find the lighter kinds of Shells, such as those of the *Echini*, and the like, very plentifully in Chalk...... Humane Bodies, the Bodies of Quadrupeds, and other Land-Animals, of Birds, of Fishes, both of the Cartilaginous, the Squamose, and Crustaceous kinds; the Bones, Teeth, Horns, and other parts of Beasts, and of Fishes: the Shells of Land-Snails: and the Shells of those River and Sea Shell-Fish that were lighter than Chalk &c. Trees, Shrubs, and all other Vegetables, and the Seeds of them: and that peculiar Terrestrial Matter whereof these consist, and out of which they are all formed,...were not precipitated till the last, and so lay above all the former, constituting the supreme or outmost *Stratum* of the Globe....The said *Strata*, whether of Stone, of Chalk, of Cole, of Earth, or whatever other Matter they consisted of, lying thus each upon other, were all originally parallel:...they were plain, caven, and regular....After some time the *Strata* were broken, on all sides of the Globe:... they were dislocated, and their Situation varied, being elevated in some places, and depressed in others ... the Agent, or force, which effected this Disruption and Dislocation of the *Strata,* was seated *within* the Earth."

Woodward's writings no doubt exercised a direct influence on the growth of our subject, but the indirect effects of his munificent bequest to the University of Cambridge and his foundation of the Chair of Geology in that University were even greater, for as will be pointed out in its proper place, two of the occupants of that chair played a considerable part in raising stratigraphical geology to the position which it now occupies.

The discoveries which were made after the publication of Woodward's book and before the appearance of the map and writings of William Smith are given in the memoir of the latter author, written by his nephew, who formerly occupied the Chair of Geology at Oxford[1]. It would appear that the fact that "the strata, considered as definitely extended masses, were arranged one upon another in a certain *settled order* or *series*" was first published by John Strachey in the *Philosophical Transactions* for 1719 and 1725. "In a section he represents, in their true order, chalk, oolites, lias, red marls and coal, and the metalliferous rocks" of Somersetshire, but confines his attention to the rocks of a limited district.

The Rev. John Michell published in the *Philosophical Transactions* for 1760 an "Essay on the Cause and Phænomena of Earthquakes," but Prof. Phillips gives proofs that Michell, who in 1762 became Woodwardian Professor, had before 1788 discovered (what he never published) the first approximate succession of the Mesozoic rocks, in the district extending from Yorkshire to the country about Cambridge. Michell's account was dis-

---

[1] *Memoirs of William Smith, LL.D.* By J. Phillips, F.R.S., F.G.S. 1844.

covered written by Smeaton on the back of a letter dated 1788. The following is the succession as quoted in Phillips' memoir (p. 136):

	Yards of thickness.
" Chalk	120
Golt	50
Sand of Bedfordshire	10 to 20
Northamptonshire lime and Portland lime, lying in several strata	100
Lyas strata.................................	78 to 100
Sand of Newark............................	about 30
Red Clay of Tuxford, and several	100
Sherwood Forest pebbles and gravel ...	50 unequal
Very fine white sand	uncertain
Roche Abbey and Brotherton limes......	100
Coal strata of Yorkshire	—"

The order of succession of the Cretaceous, Jurassic, Triassic and Permian beds will be readily recognised as indicated in this section, though the discovery of the detailed succession of the Jurassic rocks was reserved for Smith.

In the year 1778, John Whitehurst published *An Inquiry into the Original State and Formation of the Earth*, containing an Appendix in which the general succession of the strata of Derbyshire is noted. The main points of interest are that the author clearly recognised the 'toadstones' of Derbyshire as igneous rocks, "as much a *lava* as that which flows from Hecla, Vesuvius, or Ætna," though he believed that they were intrusive and not contemporaneous, and he also foreshadows the distinction between the solid strata and the superficial deposits,—" we may conclude," he says, " that all beds of sand and gravel are assemblages of adventitious bodies and not original *strata*: therefore wherever sand or gravel form the surface of the earth, they conceal the original *strata* from our observation, and deprive us of the advantages of judging, whether coal or limestone are contained in the lower regions of the earth, and more especially in flat countries where the *strata* do not basset."

Werner, who was born in 1750, exercised more influence by his teaching than by his writings. His ideas of stratigraphical geology were somewhat vitiated by his theoretical views concerning the deposition of sediment from a universal ocean, in a definite order, beginning with granite, followed by gneiss, schists, serpentines, porphyries and traps, and lastly ordinary sediments. He recognised and taught that these rocks had a definite order " in which the remains of living bodies are successively accumulated, in an order not less determinate than that of the rocks which contain them[1]." The limited value of Werner's stratigraphical teaching is accounted for by Lyell, who remarks that " Werner had not travelled to distant countries ; he had merely explored a small portion of Germany, and conceived and persuaded others to believe that the whole surface of our planet, and all the mountain-chains in the world, were made after the model of his own province," and the author of the *Principles* justly calls attention to the great importance of travel to the geologist. Those who cannot travel extensively should at any rate pay special attention to the works published upon districts other than their own, and even at the present time, the writings of some British workers are apt to be marked by some of that ' insularity ' which our neighbours regard as a national characteristic.

It is now time to turn directly to the work of William

[1] Cuvier's *Eloge*.

Smith, who, of all men, exercised the most profound influence upon the study of stratigraphical geology and may indeed be regarded as the true founder of that branch of the science. The memoir of his life which was before mentioned is all too short to illustrate the methods of work which he followed, but in it we can trace his success to three things:—firstly, his 'eye for a country,' to use a phrase which is thoroughly understood by practical geologists, though it is hard to explain to others, inasmuch as it epitomises a number of qualifications of which the most important are, a ready recognition of the main geological features from some coign of vantage, an intuitive perception of what to note and what to neglect, and the power of storing up acquired information in the mind rather than the note-book, so that one may use it almost unconsciously for future work; secondly, ability to draw conclusions from his observations, and thirdly, and perhaps most important of all in its ultimate results, a facility for checking these conclusions by means of further observations, and dropping those which were clearly erroneous, whilst extracting the truth from those which contained a germ of truth mixed with error.

Besides writers referred to above "some foreign writers, in particular Scilla and Rouelle, appear to have made very just comparisons of the natural associations of fossil shells, corals, &c. in the earth, with the groups of similar objects as they are found in the sea, and thus to have produced new proofs of the organic origin of these fossil bodies; but they give no sign of any knowledge of the *limitation of particular tribes of organic remains to particular strata*, of the *successive existence of different groups of organization*, on *successive beds of the antient sea*. Mr Smith's claim to this happy and fertile induction is clear and unquestionable[1]." We get a clue to the manner in which he arrived at his view in the following passage[2]:— "Accustomed to view the surfaces of the several strata which are met with near Bath uncovered in large breadths at once, Mr Smith saw with the distinctness of certainty, that 'each stratum had been in succession the bed of the sea'; finding in several of these strata abundance of the exuviae of marine animals, he concluded that these animals had lived and died during the period of time which elapsed between the formation of the stratum below and the stratum above, at or near the places where now they are imbedded; and observing that in the successively-deposited strata the organic remains were of different forms and structures—Gryphites in the lias, Trigoniæ in the inferior oolite, hooked oysters in the fuller's earth,—and finding these facts repeated in other districts, he inferred that each of the separate periods occupied in the formation of the strata was accompanied by a peculiar series of the forms of organic life, that these forms characterized those periods, and that the different strata could be identified in different localities and otherwise doubtful cases by peculiar imbedded organic remains[3]."

William Smith seems to have recognised intuitively

[1] *Memoir of William Smith*, p. 142.

[2] *Ibid.* p. 141.

[3] The work of Smith which directly bears upon the establishment of the law of identification of strata by included organisms is published in two treatises, entitled:—

(i) *Strata identified by Organized Fossils*, 4to. (intended to comprise seven parts, of which four only were published), commenced in 1816.

(ii) *A Stratigraphical System of Organized Fossils*, compiled from the original Geological Collection deposited in the British Museum. 4to. 1817.

the truth of a law which was but dimly understood before his time,—the law of superposition, which may be thus stated : " of any two strata, the one which was originally the lower, is the older." This may appear self-evident but it was certainly not so. As the result of this recognition he established the second great stratigraphical law, with which his name will ever be linked, that strata are identifiable by their included organisms.

Before Smith's time, geological maps were lithological rather than stratigraphical, they represented the different kinds of rocks seen upon the surface without regard to their age; since Smith revolutionised geology, the maps of a country composed largely of stratified rocks are essentially stratigraphical, but partly no doubt on account of adherence to old custom, partly on economic grounds, the majority of our stratigraphical maps are lithological rather than palæontological, that is the subdivisions of the strata represented upon the map are chosen rather on account of lithological peculiarities than because of the variations in their enclosed organisms. It is hardly likely that Government surveys will be allowed to publish palæontological maps, which will be almost exclusively of theoretical interest, and it remains for zealous private individuals to accomplish the production of such maps. When they are produced, a comparison of stratigraphical maps founded on lithological and palæontological considerations will furnish results of extreme scientific interest.

Turning now from Smith's contributions to the science as a whole, we may now consider what he did for British geology. His geological map was published in 1815 and was described as follows:—"A Geological Map of England and Wales, with part of Scotland; exhibiting the Collieries, Mines, and Canals, the Marshes and Fen Lands originally

overflowed by the Sea, and the varieties of Soil, according
to the variations of the Substrata; illustrated by the most
descriptive Names of Places and of Local Districts;
showing also the Rivers, Sites of Parks, and principal
Seats of the Nobility and Gentry, and the opposite Coast
of France. By William Smith, Mineral Surveyor." The
map was originally on the scale of five miles to an inch.
In 1819 a reduced map was published, and in later years
a series of county maps. He also published several
geological sections, including one (in 1819) showing the
strata from London to Snowdon.

The student should compare Smith's map of the
strata with one published in modern times in order to
see how accurate was Smith's delineation of the outcrop
of the later deposits of our island.

The following table, taken from Phillips' memoir,
p. 146, is also of interest as showing the development
of Smith's work and the completeness of his classification
in his later years, and as illustrating how much we are
indebted to Smith for our present nomenclature, so much
so that as Prof. Sedgwick remarked when presenting the
first Wollaston Medal of the Geological Society to Smith,
" If in the pride of our present strength, we were disposed
to forget our origin, our very speech would bewray us: for
we use the language which he taught us in the infancy
of our science. If we, by our united efforts, are chiselling
the ornaments and slowly raising up the pinnacles of
one of the temples of nature, it was he who gave the
plan, and laid the foundations, and erected a portion of
the solid walls by the unassisted labour of his hands."[1]

[1] The reader may consult an interesting paper by Professor Judd, on
"William Smith's Manuscript Maps," *Geological Magazine*, Decade IV.
vol. IV. (1897) p. 439.

COMPARATIVE VIEW OF THE NAMES AND SUCCESSION OF THE STRATA.

Table drawn up in 1799.	Table accompanying the map, drawn up in 1812.	Improved table drawn up in 1815 and 1816 after the first copies of the map had been issued.
	London Clay	1 London Clay
	Clay or Brick-earth	2 Sand
		3 Crag
	Sand or light loam	4 Sand
1 Chalk	Chalk	5 Chalk { Upper / Lower
2 Sand	Green Sand	6 Green Sand
	Blue Marl	7 Brick Earth
	Purbeck Stone, Kentish Rag and Limestone of the vales of Pickering and Aylesbury, Iron Sand and Carstone	8 Sand / 9 Portland Rock / 10 Sand / 11 Oaktree Clay / 12 Coral Rag and Pisolite / 13 Sand
3 Clay	Dark Blue Shale	14 Clunch Clay and Shale
		15 Kelloway's Stone
	Cornbrash	16 Cornbrash
4 Sand and Stone		17 Sand and Sandstone
5 Clay		
6 Forest Marble	Forest Marble Rock	18 Forest Marble
		19 Clay over Upper Oolite
7 Freestone	Great Oolite Rock	20 Upper Oolite
8 Blue Clay		
9 Yellow Clay		
10 Fuller's Earth		21 Fuller's Earth and Rock
11 Bastard ditto and Sundries		
12 Freestone	Under Oolite	22 Under Oolite
13 Sand		23 Sand
		24 Marlstone
14 Marl Blue	Blue Marl	25 Blue Marl
15 Blue Lias	Blue Lias	26 Blue Lias
16 White Lias	White Lias	27 White Lias
17 Marlstone, Indigo and Black Marls		
18 Red Ground	Red Marl and Gypsum	28 Red Marl
19 Millstone	Magnesian Limestone Soft Sandstone	29 Redland Limestone
20 Pennant Street		
21 Grays		
22 Cliff	Coal Districts	30 Coal Measures
23 Coal		
	Derbyshire Limestone	31 Mountain Limestone
	Red and Dunstone	32 Red Rhab and Dunstone
	Killas or Slate	33 Killas
	Granite. Sienite and	34 Granite. Sienite and

The above table contains a very complete classification of the British Mesozoic rocks, one of the Tertiary strata which is less complete, and a preliminary division of the Palæozoic rocks into Permian (Redland Limestone), Carboniferous (Coal Measures and Mountain Limestone), Devonian (Red Rhab and Dunstone) and Lower Palæozoic (Killas).

Since Smith's time the main work which has been done in classification is a fuller elucidation of the sequence of the Tertiary and Palæozoic Rocks, and this we may now consider.

The Mesozoic rocks are developed in Britain under circumstances which render the application of the test of superposition comparatively simple, for the various subdivisions crop out on the surface over long distances, and the stratification is not greatly disturbed. With the Tertiary and Palæozoic Rocks it is otherwise, for some members of the former are found in isolated patches, whilst the latter have usually been much disturbed after their formation.

Commencing with the Tertiary deposits we may note that "the first deposits of this class, of which the characters were accurately determined, were those occurring in the neighbourhood of Paris, described in 1810 by MM. Cuvier and Brongniart......Strata were soon afterwards brought to light in the vicinity of London, and in Hampshire, which although dissimilar in mineral composition were justly inferred by Mr T. Webster to be of the same age as those of Paris, because the greater number of fossil shells were specifically identical[1]." It is to Lyell that we owe the establishment of a satisfactory classification of the Tertiary deposits which is the basis of

[1] Lyell, *Students' Elements of Geology*. 2nd Edition, p. 118.

later classifications. Recognising the difficulty of applying the ordinary test of superposition to deposits so
scattered as are those of Tertiary age in north-west
Europe, he in 1830, assisted by G. P. Deshayes, proposed
a classification based on the percentage of recent mollusca
in the various deposits. It may be noted, that although
this method was sufficient for the purpose, it has been
practically superseded, as the result of increase of our
knowledge of the Tertiary faunas, by the more general
method of identifying the various divisions by their
actual fossils without reference to the number of living
forms contained amongst them. The further study of
the British Tertiary rocks was largely carried on by
Joseph Prestwich, formerly Professor of Geology in the
University of Oxford.

Amongst the Palæozoic rocks, it has been seen that
the Permian, Carboniferous and some of the Devonian
beds were recognised as distinct by Smith, though a
large number of deposits now known to belong to the
last named were thrown in with other rocks as 'killas.'
The Devonian system was established and the name
given to it in 1838 by Sedgwick and Murchison, largely
owing to the palæontological researches of Lonsdale. An
attempt was subsequently made to abolish the system,
but the detailed palæontological studies of R. Etheridge
finally placed it upon a secure basis. The establishment
of the Devonian system cleared the way for the right
understanding of the Lower Palæozoic rocks, which
Sedgwick and Murchison had commenced to study before
the actual establishment of the Devonian system, and to
these workers belongs the credit of practically completing
what was begun by William Smith, namely, the establishment of the Geological Sequence of the British strata.

The controversy which unfortunately marked the early years of the study of the British Lower Palæozoic Rocks is well-nigh forgotten, and in the future the names of Sedgwick and Murchison will be handed down together, in the manner which is most fitting.

Our account of the growth of British Stratigraphical Geology is not yet complete. In 1854, Sir William Logan applied the term Laurentian to a group of rocks discovered in Canada, which occurred beneath the Lower Palæozoic Rocks. Murchison shortly afterwards claimed certain rocks in N.W. Scotland as being of generally similar age, and since then a number of geologists, most of whom are still living, have proved the occurrence of several large subdivisions of rocks in Britain, each of which is of pre-Palæozoic age.

The above is a brief description of the growth of our knowledge of the order of succession of the strata which is the foundation of Stratigraphical Geology. A sketch of the manner in which the knowledge which has been obtained has been applied to the elucidation of the earth's history of different times would require far more space than can be devoted to it in a work like the present, but some idea of it may be gained from a study of the later chapters of the book. It will suffice here to remark, that to Godwen-Austin we owe the foundation of what may be termed the physical branch of Palæophysiography, which is concerned with the restoration of the physical conditions of past ages, while Cuvier and Darwin have exerted the most influence on the study of Stratigraphical Palæontology.

CHAPTER III.

NATURE OF THE STRATIFIED ROCKS.

THE present constituents of the earth which are accessible for direct study are divisible into three parts. The inner portion, consisting of *rocks,* is known as the *lithosphere;* outside this, with portions of the lithosphere projecting through into the outermost part, is the *hydrosphere,* comprising the ocean, lakes, rivers, and all masses of water which rest upon the lithosphere in a liquid condition. The outermost envelope, which is continuous and unbroken is the *atmosphere,* in a gaseous condition. It is well known that some of the constituents of any one of these parts may be abstracted from it, and become a component of either of the others; thus the atmosphere abstracts aqueous vapour from the hydrosphere, and the lithosphere takes up water from the hydrosphere, and carbonic anhydride from the atmosphere.

The nebular hypothesis of Kant and Laplace necessitates the former existence of the present solid portions of the lithosphere in a molten condition, and accordingly the first formed solid covering of the lithosphere, if this hypothesis be true, must have been formed from molten material, or in the language of Geology, it was an *igneous rock.* Consequently, the earliest *sedimentary rock* was necessarily derived directly from an igneous rock, with possible addition of material from the early hydrosphere and atmosphere, and all subsequently formed sedimentary rocks have therefore been derived from igneous rocks (with the additions above stated) either directly, or indirectly through the breaking up of other sedimentary rocks which were themselves derived directly or indirectly from igneous rocks. The observations of geologists show that this supposition that the materials of sediments have been directly or indirectly obtained for the most part from once-molten rocks is in accordance with the observed facts, and so far their observations testify to the truth of the nebular hypothesis. This being the case, the study of the petrology of the igneous rocks is necessary, in order to arrive at a true understanding of the composition of the sedimentary ones. The igneous rocks are largely composed of four groups of minerals, viz.—quartz, felspars, ferro-magnesian minerals, and ores. Of these the quartz (composed of silica) yields particles of silica for the formation of sedimentary rocks; the felspars, which are double silicates of alumina and an alkali or alkaline earth, being prone to decomposition furnish silicate of alumina and compounds of soda, potash, lime, &c. The ferro-magnesian minerals (such as augite, hornblende and mica) may undergo a certain amount of decomposition, and yield compounds of iron, lime, &c. We may also have fragments of any of these minerals, and of the ore group in an unaltered condition. The composition of a sedimentary rock which has undergone no alteration after its formation will therefore depend upon the character of the rock from which it was derived, the chemical changes which take place in the materials which compose it, before they enter into its mass, and the mechanical sorting which they undergo prior to their deposition.

In the above passage the terms igneous rock and sedimentary rock have been used, and it is necessary to give some account of the sense in which they were used.

An *igneous* rock is one which has been *consolidated* from a state of *fusion*. It is not necessary to discuss here the exact significance of the word fusion, and whether certain rocks which are included in the igneous division were formed rather from solution at high temperature than from actual fusion. This point is of importance to the petrologist, but to the student of stratigraphical geology the term igneous rock may be used in its most comprehensive sense. These igneous rocks were consolidated either upon the surface of the lithosphere or in its interior.

The other great group of rocks is one to which it is difficult to apply a satisfactory name. They have been termed by different writers, *sedimentary*, *stratified*, *derivative*, *aqueous*, and *clastic*, but no one of these terms is strictly accurate. The term *sedimentary* implies that they have settled down, at the bottom of a sheet of water for instance. It can hardly be maintained that limestones formed by organic agency, like the limestones of coral reefs, are sedimentary in the strict sense of the term, and an accumulation like surface-soil can only be called a sediment by straining the term. *Stratified* rocks are those which are formed in strata or layers, but many of the rocks which we are considering do not show layers on a small scale, and igneous rocks (such as lava-flows) are also found in layers, though such layers are not true strata in the sense in which the term is used by geologists; the term *stratified* is perhaps the least open to objection of any of those named above. *Derivative* implies that the fragments have been derived from some pre-existing rock, but as there are many ways in which fragments of one rock may be derived from another, the term is too comprehensive. *Aqueous* rocks should be formed in water, and most of the class of rocks which we are considering have been so formed, but others such as sand-dunes and surface-soil have not. (The term Aerial or Æolian has been suggested to include these rocks which are thus separated from the Aqueous rocks proper; the objection to this is that the origin of these rocks is closely connected with that of the true Aqueous rocks, and moreover the group is too small to be raised to the dignity of a separate subdivision.) Lastly, the name *clastic* has been given, because the rocks so called are formed by the *breaking up* of pre-existing rocks. There are two objections to this name. In the first place, some rocks included under the head clastic are formed by solution of material and its consolidation from a state of solution by chemical or organic agency, though we may perhaps speak of rocks being broken up by chemical as well as by mechanical action. The most important objection is that many clastic rocks are formed by the breaking up of rocks subsequently to their formation, and it has been proposed that rocks of this nature should be termed *cataclastic*, while those which are formed by the breaking up of pre-existing rocks upon the earth's surface should be termed *epiclastic;* another group formed of materials broken up within the earth, and accumulated upon its surface as the result of ejection of fragmental material from volcanic vents being termed *pyroclastic.* This classification is scientific, and under special circumstances is extremely useful, but the older terms have been used so generally, and with this explanation their use is so unobjectionable, that they may be retained, and the term *stratified* will be generally used to indicate all rocks which are not of igneous origin or formed as mineral veins in the earth's interior.

The division of rocks into *three* great groups, the Igneous, Stratified and Metamorphic (the latter name being applied to those rocks which have undergone considerable alteration since their formation), is objectionable, since we have metamorphic igneous rocks as well as metamorphic stratified ones. The most convenient classification is as follows:—

A. Igneous $\begin{cases} 1. & \text{Unaltered.} \\ 2. & \text{Metamorphic.} \end{cases}$

B. Stratified $\begin{cases} 1. & \text{Unaltered.} \\ 2. & \text{Metamorphic.} \end{cases}$

It must be distinctly understood that all geological phenomena must be taken into account by the stratigraphical geologist. The upheaval of strata, the production of jointing and cleavage in them, their intrusion by igneous material, their metamorphism, give indications of former physical conditions equally with the lithological characters of the strata, and their fossil contents. Nevertheless it is not proposed to give a full account of the various phenomena displayed by rocks; the student is referred to Text-books of General Geology for this information. It will be as well here, however, to point out in a few words the exact significance of the existence of strata in the lithosphere.

The formation of strata and their subsequent destruction to supply material for fresh strata are due to three great classes of changes. Beginning with a portion of lithosphere composed of rock, it is found that rock is broken up by agents of denudation, as wind, rain, frost, rivers and sea. These agents perform their function mainly upon the portion of the lithosphere which projects through the hydrosphere to form *land*, and the land is the main area of denudation. The materials furnished by denudation are carried away, and owing to gravitation, naturally proceed from a higher to a lower level, often resting on the way, but if nothing else occurs, ultimately finding their way to the *sea*, where they are deposited as strata. The sea is the principal area for the reception of this material, and it is there accordingly that the bulk of stratified rock is formed. If nothing else occurred, in time the whole of the land would be destroyed, and the wreckage of the land deposited beneath the sea as stratified rock. As it is there is a third class of change, underground change, causing movements of the earth's crust (to use a term which can hardly be defined in few words but which is generally understood), and as the result of the relative uplift of portions of the earth's crust, the stratified rocks formed beneath the oceans are raised above its level, giving rise to new masses of land, which are once more ready for destruction by the agents of denudation. This cycle of change (all parts of which are ever proceeding simultaneously) is one of the utmost importance to the stratigraphical geologist.

Stratification is the rock-structure of prime importance in stratigraphical geology, and a few words must here be devoted to its consideration, leaving further details to be dealt with hereafter. The surface of the ocean-floor is, when viewed on a large scale, so level, that it may be considered practically horizontal, and accordingly in most places the materials which are laid down on the ocean-floor give rise to accumulations which at all times have a general horizontal surface (when the ocean-slopes depart markedly from horizontality the deposits tend to abut against these slopes rather than to lie with their upper surfaces parallel to their original angle). A practically horizontal surface of this character may give rise to a *plane of stratification* (or *bedding-plane*) in more than one way. A pause may occur during which there is a cessation of the supply of material, so that the material which has already been accumulated has sufficient time to become partially consolidated before the deposition of fresh material upon it. In this way a want of coherence between the two masses is produced, along the plane of junction, which after consolidation of the deposits causes an actual divisional plane along which the two deposits may be separated. This is a plane of stratification. The pause may be produced in various ways, sometimes between successive high tides, at others as the result of physical changes which may have taken ages to happen. Again, after material of one kind has been deposited, say sand, some other substance such as clay may be accumulated on its upper surface, giving rise to a plane of stratification between two deposits of different lithological characters. If this occurs alone, there may be actual coherence between the two strata, so that it is erroneous to speak of a plane of stratification as if it were always one along which one deposit could be readily split from the other, though as a general though by no means universal rule, change from one kind of deposit to another is also marked by want of coherence between the two. The material between two planes of stratification forms a *stratum* or *bed*, though if the deposit be very thin it is known as a *lamina*, and the planes are spoken of as *planes of lamination* (no hard and fast line can be drawn between strata and laminæ; several of the latter usually occur in the space of an inch).

A *stratum* will have its upper and lower surface apparently parallel, though not really so, for no stratum extends universally round the earth, and many of them disappear at no great distance when traced in any direction. Parts of one stratum may be composed of different materials from other parts when traced laterally, thus one stratum may be found composed essentially of sand in one place, of mud in another, and of a mixture of the two in an intervening locality. Whatever be the composition of a stratum it dies out eventually, owing to the coming together of the upper and lower bounding planes of stratification. The stratum is thickest at some spot, from that spot it becomes thinner in all directions, until it disappears at last by the coalescence of the bounding-planes. This is spoken of as *thinning-out*. Strata, then, consist of lenticular masses of rock, separated from the underlying and overlying strata by planes of stratification. The shape of the lenticle may vary immensely, the thickness bearing no definite relationship to the horizontal extent. Some strata, many feet in thickness, may thin out and disappear completely in the course of a few yards, whilst others an inch or two in thickness may be traced horizontally for many miles. We often find thin strata of coal and limestone, extending for great distances, strata of mud thinning out more rapidly, and sandstones still more rapidly, but no universal rule connecting rapidity of thinning-out with composition of the strata can be laid down.

Having seen what a stratum is, it now remains to speak of the composition of the stratified rocks. They have been classified according to their composition, and according to their origin. According to composition they have been divided into:

Arenaceous rocks, composed essentially of grains of sand.

Argillaceous rocks, composed essentially of particles of mud.

Calcareous rocks, composed essentially of particles of carbonate of lime.

Carbonaceous rocks, composed largely of hydrocarbon compounds.

Siliceous rocks, composed essentially of silica not in the form of grains;

whilst according to their origin they have been separated into :—

Mechanically-formed rocks, composed of fragments derived from other rocks by mechanical fracture.

Chemically-formed rocks, composed of particles which have been chemically deposited from a state of solution.

Organically-formed rocks, composed of materials which have been derived from a state of solution or from the gaseous condition by the agency of organisms.

Whichever classification be adopted (and each is useful for special purposes), it must be noted that no hard and fast line can be drawn between one division and another. A rock may be partly arenaceous and partly calcareous, composed of a mixture of sand and lime, and the same rock may similarly be partly mechanically and partly organically formed, the sand being due to mechanical fracture, and the lime to the agency of organisms, and so with the other divisions.

As many of the changes which have occurred in past times have been concerned in destruction and obliteration, whilst deposition is the cause of preservation, the study of deposits is peculiarly adapted for testing the truth of the grand principle of geology that the changes which have taken place in past times are generally speaking similar in kind and in intensity of action to those which are in progress at the present day, and a study of the modern deposits is specially important as throwing light upon the characters of those which have been formed in past times. It will be abundantly shown in the sequel that the deposits of the strata are in general comparable in all essential respects with those which are being formed at present, and accordingly they give most valuable indications as to the nature of the physical and other conditions under which they were laid down. The desert sand, the precipitate of the inland sea, the reef-limestone and many another deposit can thus be detected by an examination of their lithological characters, combined with consideration of other kinds of evidence. The petrology of the sedimentary rocks is still in its infancy, though much has already been done, but it offers a wide field of inquiry to the field-geologist and worker with the microscope [1].

[1] The student will do well to consult *The Challenger Report* by Messrs Murray and Renard (1891), for information concerning many modern sediments, and Harker's *Petrology for Students* Section D, for general information on the Petrology of the Stratified Rocks.

CHAPTER IV.

THE LAW OF SUPERPOSITION.

IN a previous chapter this law was given as follows: "Of any two strata, the one which was originally the lower is the older;" the general truth of the law depends upon the fact that except under very exceptional circumstances the strata are deposited upon the surface of the lithosphere, and not beneath it. There are occasions where strata may be deposited beneath the lithosphere, but as a general rule the geologist will not be misled by such occurrences. In caverns, accumulations often occur which are newer than the strata over the cavern roof, and so long as caverns are formed in ordinary sedimentary rocks, no great difficulty will result from this exception to the law of superposition. When caverns occur beneath masses of land ice, the order of superposition may be misleading. A deposit may be formed on the surface of the ice, and subsequently to this a newer deposit may be laid down in a sub-glacial or englacial cavern; upon the melting of the ice the newer deposit would be found with the older one resting upon its surface.

Apart from these exceptional cases, the law as stated holds good, but the reader will notice the insertion of the word 'originally' which requires some comment.

A geologist speaks of one bed lying *upon* another not only when the beds are horizontal, but when they are inclined at any angle, until they become vertical, so that until beds have been turned through an angle of 90° by earth-movement the test of superposition is applicable, but when they have been turned more than 90°, the stratum which was originally lower rests upon that which was originally above it, and in the case of these *inverted* strata, the test of superposition is no longer applicable. It was formerly supposed that cases of inversion were comparatively rare and local, and that the test of super-position could therefore be generally applied with confidence, but it is now known that though this is generally true of such strata as have been subjected only to those wide-spread, fairly uniform movements which are spoken of as *epeirogenic* or continent-forming, where the radius of each curve is very long, inversion is a frequent accompaniment of the more local *orogenic* or mountain-forming movements, where the radius of a curve is short. Though orogenic movements are limited as compared with those of epeirogenic character, they often affect large tracts of country, in which case the apparent order of succession of the strata need not be the true one, and examples of inversion may be frequent [1].

It is not easy to lay down any definite rules for detecting inverted strata, where the top of an inverted arch is swept off by denudation or the bottom of an inverted trough concealed beneath the surface, beyond stating that if an easily recognised set of beds is obviously

[1] For a discussion of the principles of mountain-building see Heim, A., *Untersuchungen über den Mechanismus der Gebirgsbildung*, and Lapworth, C. "The Secret of the Highlands," *Geological Magazine*, Decade II. vol. x. pp. 120, 193, 337.

repeated in inverse order, inversion must have occurred, though even then it may not be clear which side of the fold shows the beds in original and which in inverted sequence. Suggestions are frequently made that ripple-marks and worm-tracks may be utilised in order to discover inversion, for the well-formed ripple-marks will appear convex on the upper surface of a bed which is not inverted, and we may note concave casts of these ripple-marks on the under surface of the overlying bed, whilst worm-tracks will appear concave on the upper surface, and their casts convex on the lower surface of the succeeding bed under similar conditions. In the case of inversion the occurrences will be the exact opposite to these. Unfortunately ripple-marks and worm-tracks may, as will appear in the sequel, be simulated by structures produced in quite a different way, and unless the observer is certain that he is confronted with true ripple-marks and worm-tracks he may be seriously misled. The geologist must take into account all the evidence at his disposal, when he is dealing with cases of possible inversion, but oftentimes he will after due consideration of all the phenomena be left in doubt unless he is able to supplement his observations on the succession of the strata by evidence derived from the included fossils.

The test of superposition is most apt to be misleading when the strata have been affected by the faults known as reversed faults or thrust-planes.

Reference to text-books will show that a fold consists of two parts, the arch and the trough, and that the two are connected by a common-, middle-, or partition-limb. In the case of an inverted fold, an **S**-shaped or sigmoidal structure is the result (Fig. 1 A).

Here the portions of any bed (*xx*) which occur in the

arch or trough are in normal position, and have not been moved round through an angle of 90°, whilst the portion which occurs in the common limb *c* has been moved round through an angle greater than 90° and is inverted, so that its former upper surface now faces downwards. In Fig. 1 B the common limb is replaced by a reversed fault, or thrust-plane, and the inverted portion of the bed seen in the common limb is therefore absent. An observer, applying the test of superposition, might suppose that the position of the bed *x* on the left-hand side of the figure

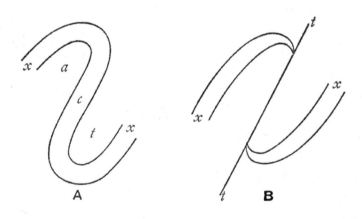

Fig. 1.

A. A sigmoidal fold, showing a bed *xx* in an overfold with arch (*a*), trough (*t*) and common limb *c*.

B. A similar bed *xx* affected by a thrust-plane *tt* which replaces the common limb.

was a different bed from the portion which is seen on the right-hand side, instead of belonging to the same bed, and in this way, if a number of parallel thrust-planes affected one bed or a set of beds, he might be led to infer the occurrence of a great thickness of strata where there was in reality a slight thickness, or even one bed only repeated again and again by faulting. It is quite certain that exaggerated estimates of the thickness of deposits have frequently been made owing to the non-recognition of the occurrence of repetition as a consequence of the existence of thrust-planes.

Where thrust-planes are suspected, it is well to look for some of the following features:

(a) The strata of a country affected by thrust-planes often crop out as lenticular masses, thinning out rapidly in the direction of the strike[1]. This is true of beds thrown into sharp folds whether or not inverted, but the lenticles will be wider in a direction at right angles to that of the strike as compared with their length when inversion has not occurred. It is also true of beds which were originally deposited as lenticles, such as many massive sandstones, and as almost any kind of deposit may be formed originally as a lenticle, the test by itself is by no means sufficient as a proof of thrusting, though it is suggestive.

(b) The *surfaces* of the strata are often affected by the striations known as slickensides, and the joint-faces of gently inclined beds are also frequently marked by slickensides which often run in a nearly horizontal direction.

(c) A parallel structure presenting the appearances characteristic of the mechanically-formed features of a foliated rock is often developed, and one or more of certain accompanying phenomena will probably be found, which will be noticed more fully in a later chapter.

(d) Extension or stretching of the rocks will have been frequently produced, causing rupture, and the resulting fissures are usually filled with mineral-veins, though

[1] For definitions of the terms dip, strike, outcrop and allied expressions, the reader is referred to a *Text-Book of Geology*.

this occurrence is by no means characteristic of rock
which have been affected by thrust-planes.

(e) Chemical changes may have occurred which have
resulted in the reconstitution of some of the rock-consti
tuents, which may crystallise where pressure is least, thu
we often find rocks which have undergone movements o
the type we are considering marked by the existence o
sericitic films upon the surfaces.

Another reservation must be made when considering
the law of superposition. The test is only applicable fo
limited areas. Suppose we find a deposit of clay *a* restin
upon another deposit of limestone *b* in the south o
England, and can prove that the apparent succession i
the true one, that is, that there has been no inversion; i
is clear that the test of superposition is applicable in tha
area. Now, we may be able to trace the two deposit
continuously across the country, one as a clay, the othe
as a limestone; so that when we reach the north c
England we find the clay *a* still reposing upon the lime
stone *b*. The test of superposition is applicable in tha
area also, the clay of the northern area being newer tha
the limestone of the same region. But, for reasons whic
will ultimately appear, it by no means follows that th
clay of the north is newer than the limestone of the soutl
although the two deposits are continuously traceable wit
the same lithological characters; it may have been forme
simultaneously with the limestone of the south, or eve
before it. Something more, therefore, than the test
superposition is necessary in order to make out tl
relative ages of continuous deposits in a wide region, ar
this is still truer in the case of deposits which are di
continuous, whether separated from one another by tl
sea, or by outcrops of older or newer rocks.

A few words of warning may be added with reference to the detection of bedding-planes. A bedding-plane is one which separates two beds, and its existence is determined during the deposition of the beds. Many other planes are formed in rocks subsequently to their deposition, and it is not always easy to distinguish these from true bedding-planes. That even experienced observers may be led astray is shown by the fact that, of recent years, it has been proved that great masses of rock have been claimed as of sedimentary origin, and their apparent order of succession noted, which are in truth naught but irregular masses of intrusive igneous rocks affected by divisional planes which simulate bedding, produced in the rocks subsequently to their consolidation. Joints, faults, and cleavage-planes may all at times simulate planes of bedding, and it is frequently very difficult to distinguish them in the limited exposures with which a geologist has oftentimes to deal. It is easier to make suggestions for distinguishing bedding-planes from other planes which simulate them, than to apply the suggestions in practice, and the student of field geology will find that experience is the only guide, though after years of experience he may be confronted with cases where the evidence is insufficient to convince him that he is dealing with planes of stratification and not with some other structure.

From what has been remarked, it will be inferred that the test of superposition though of prime importance to the geologist is frequently insufficient to enable him to ascertain the true order of succession of the strata, and he is compelled to supplement this test by some other. There are several useful physical tests which may frequently be applied. Thus, if a rock *a* contains fragments of another rock *b, under such circumstances as to show that*

the fragments of b *were included in* a *during its deposition*
it is clear that *b* is older than *a.*   Here again, it will b
found from what appears in a later chapter that th
student is confronted with difficulties when actuall
examining rocks, for fragmental rocks of cataclastic origir
where the fragments have been formed as the result o
fracture produced by earth-movements subsequently t
the deposition of the rock, simulate epiclastic rocks i
which the fragments were introduced during the accumu
lation of the deposits to so surprising a degree as sometime
to baffle the most experienced observer.   Not only ar
the fragments of these cataclastic rocks broken up, bu
they may be further rounded so as to imitate in a remark
able manner the waterworn pebbles of an epiclasti
conglomerate.   Again, an older series of rocks may hav
had structures impressed upon them as the result o
changes subsequent to their formation, and before th
formation of a newer set which the latter therefore do no
exhibit.   Joining, cleavage, and various metamorphi
phenomena may thus be exhibited by the older rocks
but great care is required in applying this test, especiall
with a limited thickness of rocks, as one set may no
exhibit the structures not because they were not in ex
istence when the structures were developed, but becaus
their nature is such that they were incapable of receivin
or retaining the structures.   For instance a mass of gri
which is older than a mass of clay-slate may not b
cleaved, because, although subjected to the pressure whic
produced the cleavage, it was of a nature not adapted t
the development of cleavage structure.

On the whole, application of tests dependent upo
physical features of rocks, does not often supplement t
any great extent the information supplied by ascertainin

the order of superposition, and in all cases, where possible, every other kind of information should be supplemented, by that which is acquired after examination of the included organisms of the strata.

CHAPTER V.

THE TEST OF INCLUDED ORGANISMS.

THE second great law of the Stratigraphical Geologist is that fossiliferous strata are identifiable by their included organisms, in other words, that we can tell the geological age of deposits by examination of the fossils contained in them, though the determination of age must be given in more general terms in some cases than in others. Con· siderable misconception has arisen concerning the value of fossils as indices of age, and it is necessary therefore to discuss the significance of the law of identification of strata by their included organisms at some length.

The comparison between fossils and medals has fre· quently been made and fossils have well been styled the "Medals of Creation"; and the significance of fossils as guides to the age of deposits may perhaps be made clearer if we pursue this comparison some way. In the first place there is clear indication of a gradual increase in the complexity of organisation of the fossils as one passes from the earlier to the later rocks, and accordingly the general facies of a fauna is likely to furnish a clue to the age of the rocks in which it is found, even though every species or even genus represented in the fauna was previously unknown to science. So an antiquary versed in the evolution of art or metallurgy, might detect the general age of a medal with whose image and superscription he was not acquainted. He would know that a medal struck in iron was formed subsequently to the bronze age, or that one formed of palladium appertained to the present century. But quite apart from any theoretical knowledge, an antiquary would find as the result of accumulated experience that certain medals are characteristic of certain periods; he would learn that the denarius is characteristic of a different period from that indicated by the coin of the Victorian era, even though he had no knowledge of the technicalities of numismatics. The same is the case with the geologist. He may not be influenced by any knowledge of the evolution of faunas and floras, but actual work amongst the rocks will show him that the trilobite is not found with the belemnite or the ichthyosaur with the elephant, save under exceptional circumstances, which only prove the rule, as for instance when worn bones of ichthyosaurs are washed from their proper strata into gravels with elephant-bones.

It must be distinctly understood that the determination of fossils as characteristic of different periods is solely made as the result of experience. No à priori reasoning may give one indication of the actual range in time of a species or genus; no one can say why *Discina* has a long range in time, whilst that of the closely related *Trematis* is very limited. This being the case, the greater the mass of evidence which is accumulated as to the range of a fossil, the greater will be the value of that fossil as a clue to the age of the deposit in which it is found. This is so important, that it requires more than mere notice. If a fossil is found in abundance in a group of strata B in any one area, and is not found in an underlying group A or overlying group C in that area after prolonged search, we may confidently speak of the fossil as characteristic of the strata B in that area, though without further work, the value of the fossil as a clue to age in other areas would be unproved. It may nevertheless happen, that after more prolonged search in A or C, in the original area a few specimens of the fossil which has been spoken of as characteristic of B may be found in one or other of them, in small quantity. The value of the fossil as one characteristic of B will be slightly diminished, though only slightly, as it is not likely to turn up in numbers in the strata A or C after the prolonged search. Should the fossil be found also to be characteristic of the strata B in areas other than the original one, it becomes of more than local value, and if, after much study it is found to characterise the same strata over wide areas, the cumulative evidence now obtained will render the fossil peculiarly important to the stratigraphical geologist. The detection of characteristic fossils is not quite so simple as might be supposed from the above remarks, for examination of the position of one fossil will not prove the contemporaneity of beds in different places, to prove this, all the evidence at our disposal must be considered, for reasons which will be presently pointed out.

As the result of accumulated knowledge, we can now compile lists of characteristic fossils of the major subdivisions of the strata, which are of world-wide utility and as our knowledge increases, we are enabled to subdivide the strata into minor divisions of more than local value.

What is a fossil? Before discussing the value of fossils as aids to the stratigraphical geologist, it may be well to make a few observations as to what constitutes a fossil. It is difficult to give any concise definition, and as is often the case in geology, an explanatory paragraph is of more value than a mere definition. The term fossil was originally applied to anything dug up from the rocks of the earth's crust, and was used with reference to inorganic objects as well as organic remains, for instance minerals were spoken of as fossils. It is now applied essentially though not exclusively to relics of former organisms, though one still reads of fossil rain-drops, fossil sun-cracks, and so on. Furthermore, the relics need not necessarily be parts of the organism, the track of a worm or a bird's nest if embedded in the strata would be termed a fossil. It is generally agreed that no sharp line can be drawn between recent and fossil organic remains which is based upon the degree of mineralisation (or as it was sometimes termed petrifaction) of the relics, for many true fossils have not undergone mineralisation, subsequent to their entombment.

It has been suggested that the name fossil should be applied to organic remains which have been entombed by some process other than human agency, but this restriction is undesirable. The stone-implement of the river gravels is as genuine a fossil as the ammonite extracted from the chalk, and the human relics of very recent date may give information of a character quite similar to that supplied by other remains, for instance, the occurrence of moa-bones in New Zealand in accumulations below those containing biscuit-tins and jam-pots has been used as a geological argument pointing to the extinction of the moa before the arrival of Europeans in New Zealand. The biscuit-tin here serves all the purposes of a fossil, and there is no valid reason why it should not be spoken of as such.

This statement brings one to consider another method which has been adopted in order to separate fossil

organisms from recent ones, namely the time-test. This
again is inapplicable, for no line can be drawn between the
shell which was buried in yesterday's tidal deposit and
that which has lain in the strata through geological ages,
and each may be equally useful to the geologist.

Whilst, then, we can give no definition of fossil which
is likely to meet with general acceptance, the term can
be so used, as not to give rise to any doubts as to its
meaning, and it is generally applicable to any organic
relics which have been embedded in any deposit or accumulation by any agent human or otherwise.

Mode of occurrence of fossils. It will not be out of
place to say a few words as to the way in which fossils
are found in strata, as beds are often inferred to be unfossiliferous, because of ignorance of methods which should
be pursued in searching for organic relics. It is unnecessary to dilate upon the actual modes of preservation
of organisms, which is treated of fully in other works. In
the first place, it is rash to assert that any deposit is unfossiliferous because no fossils have been found in it, even
after prolonged search. The Llanberis slates had been
eagerly searched for fossils for many years without result,
but that the search was not exhaustive was proved by the
discovery of trilobites in them some years ago. Seekers
after fossils are rather prone to confine their attention to
strata which are already known to be fossiliferous than to
pay much attention to those which have hitherto yielded
no organic remains.

Some kinds of deposits are more often fossiliferous
than others. Limestones which are frequently largely of
organic origin, are often rich in remains, and muddy
deposits more frequently furnish fossils than those of a
purely sandy nature. The difference in the yield is not

necessarily due to the original inclusion of more remains in one rock than in another, but is often caused by the obliteration of former relics owing to changes which have taken place in the rocks subsequently to their deposition. No sedimentary rock must be regarded as unfossiliferous, however unfitted it appears for the preservation of fossils. The writer has seen fossils, not only in coarse conglomerates, rocks which frequently contain no traces of organisms, but in deposits composed largely of specular iron ore, and even in intrusive igneous rocks, though in the latter case, the inclusion of fossils was due to circumstances which cannot have occurred with frequency.

In sandy strata, the substance of the fossils has often been completely removed, leaving hollow casts, which may be almost or quite unrecognisable. In these circumstances, much information may be obtained by taking impressions of the casts in modelling wax or some other material. The importance of this process may be judged from the results it yielded to Mr Clement Reid in the case of the fossils of the Pliocene deposits occurring in pipe-like hollows in the Cretaceous rocks of Kent and the discovery of the remarkable reptiles described by Mr E. T. Newton from the Triassic sandstones of Elgin.

In argillaceous rocks which have been affected by the processes producing cleavage, the fossils may be distorted beyond recognition or owing to the difficulty of breaking the rocks along the original planes of deposition, may remain invisible. Under such circumstances, small nodules of sandy or calcareous nature may sometimes be found included in the argillaceous deposits and may perhaps yield fossils. Oftentimes, also, where the argillaceous rock is in close proximity to a harder rock, such as massive grit, the argillaceous rock in close contiguity to the hard rock may escape the impress of cleavage-structure, and fossils may be readily extracted from rocks in this position when not obtainable from other parts of the deposit. It was under these circumstances that the trilobites alluded to above were obtained from the Llanberis slates.

The fossils of calcareous rocks are often very obvious, but difficult to extract, as they break across when the rock is fractured. They are frequently obtainable in a perfect condition when the rock is weathered. Occasionally they may be extracted from certain argillaceous limestones if the limestone be heated to redness, and suddenly plunged into cold water. Fossils are often found in a state which enables them to be readily extracted when a limestone is coarsely crystalline, though they cannot be extracted in a perfect condition when the same limestone is in a different state.

Many microzoa, which are invisible in rocks, even when viewed through a lens, may be found in microscopic sections of calcareous and silicious rocks, and plant structures may be detected under similar circumstances in the case of carbonaceous rocks.

Various special methods of extracting fossils from rocks have been described by different writers, many of which are very complex, and require much time. The mechanical action of the sand-blast and the solvent action of various acids as hydrochloric and hydrofluosilicic have been found of use upon different occasions[1]. The various processes which have been utilised in order to extract and develop fossils can, however, be best learned by information obtainable from curators of palæontological collections,

---

[1] For information concerning use of acids see especially Wiman, C. "Ueber die Graptoliten," *Bull. Geol. Inst.*, Upsala, No. 4, vol. II. Part II.

and by actual experience, and there is yet much information to be acquired as to the manner of extracting fossils
from various kinds of rocks.

Relative value of fossils to the Stratigraphical Geologist.
It has been hinted above that no general rule as to the
relative value of fossils as guides to the age of strata
can be laid down, and that the ascertainment of their
relative value is largely the result of actual experience.
It may be noted, however, that organisms which possess
hard parts are naturally more important to the geologist
than those which do not, as few traces of the latter are
preserved in the fossil state, and even when preserved
are usually too obscure to be of much practical use.
Of the organisms which do possess hard parts, different
groups have been utilised to a different degree, and one
group will be more or less important than another,
according to the use to which it is applied. Groups of
organisms which have a long range in time are naturally
useful for the identification of large subdivisions of the
strata, whilst those which have had a shorter range are
valuable when separating minor subdivisions.

Again, as the bulk of the sedimentary deposits has
been formed beneath the waters of the ocean, relics of
marine organisms are naturally more useful than those
of freshwater ones. Other things being equal, the more
easily the organism is recognisable, and the more abundant
are its remains, the greater its value to the stratigraphical
geologist, and as the remains of invertebrates are usually
found in greater quantities and in more readily recognisable condition than those of the vertebrates, they have
been used more extensively as indices of age. Of the
invertebrates, the mollusca are often very abundant,
their remains are adapted for preservation, and their

characteristics have been extensively studied, and accordingly they have been and are of great use to the geologist. Of other groups, the graptolites, corals, echinids, brachiopods, and trilobites have been very largely utilised. The Lower Palæozoic strata have been divided into numerous groups, each characterised by definite forms of graptolites, and a similar use has been made of the ammonites in the case of the Mesozoic rocks. It is not to be inferred that these groups of organisms are naturally more useful than other groups, on account of the extent to which they have been used; we can merely state that they have been proved to be useful as the result of prolonged study; when other groups have received equal attention, they may well be found to be equally useful for the purposes which we have in view.

Contemporaneity and Homotaxis. From what has been already stated, it will be recognised that the ages of the various fossiliferous rocks of the geological column[1] in any one area can be identified with greater or less degree of certainty by reference to their included organisms, the various subdivisions being marked by the possession of characteristic fossils, and it will be naturally and rightly inferred that the greater the number of characteristic fossils of any one deposit, the more certain is the identification of that deposit. In practice, geologists are wont to ascertain the age of the strata after consideration of all the fossils found therein, some of which may be actually characteristic whilst many may come up

[1] Although the rocks do not always lie on one another in regular succession, it is often convenient to speak of them as though they did, and as though a column of strata could be carved out in any region consisting of horizontal bands of deposit one above another. We speak of such an ideal arrangement as constituting a 'geological column.'

from the strata below, or pass into those above. Having ascertained the order of succession and fossil contents of the strata in various regions, it is the task of the geologist to compare the strata of these two regions, and this task is fraught with considerable difficulty. Much controversy has arisen as to the degree of accuracy with which strata of remote regions can be correlated, and the subject is one which requires full consideration.

Suppose that a series of strata which we will call A, B, and C is found in any one area, each member of which contains characteristic fossils which enable it to be recognised in that area, and we will further suppose that in another area a series of strata A', B', and C' is discovered, of which A' has the fauna of A in the former area, and similarly B' the fauna of B, and C' that of C.

It cannot be assumed that the stratum A is therefore contemporaneous with A', B with B', and C with C', but on the other hand, it must not be assumed that they are not contemporaneous. This is a statement which requires some comment. It has been urged that if the deposits A and A' in different localities contain the same fauna, this is a proof that the two are not contemporaneous, for some time must have elapsed in order to allow of the migration of the organisms from one area to another, it being justifiably assumed that they did not originate simultaneously in the two areas. But everything depends on the time taken for migration as compared with the period of existence of the fauna. If the former was extremely short as compared with the latter it may be practically ignored, for we might then speak of the strata as contemporaneous, just as a historian would rightly speak of events in the same way which occurred upon the same afternoon, though one might have happened an hour before the other. Let us then glance at the evidence which we have at our disposal, which bears upon this matter.

The objection to identification of strata with similar faunas as contemporaneous was urged by Whewell, Herbert Spencer, and Huxley, and the latter suggested the term Homotaxis or similarity of arrangement as applicable to groups of strata in different areas, in which a similar succession of faunas was traceable, maintaining that though not contemporaneous the strata might be spoken of as homotaxial. Huxley went so far as to assert that "for anything that geology or palæontology are able to show to the contrary, a Devonian fauna and flora in the British Islands may have been contemporaneous with Silurian life in North America, and with a Carboniferous fauna and flora in Africa[1]," a statement which few if any living geologists will endorse. If the statement be true, and the fauna which we speak of as Devonian, when present be always found (as it is) above that which we in Britain know as Silurian and below that which we term Carboniferous, the faunas must have originated independently in the three centres, and disappeared before the appearance of the next fauna, or having originated at the same centre, each must have migrated in the same direction, spread over the world, and become extinct as it reached the point or line from which it started. Suppose for instance a fauna A originates at the meridian of Greenwich, migrates eastward, and dies out again when it once more reaches Greenwich, that B and C do the same, at a later period, then the fauna B will always be found above A and C

---

[1] Huxley, T. H. "Geological Contemporaneity and Persistent Types of Life," being the Anniversary Address to the Geological Society for 1862 ; reprinted in *Lay Sermons, Addresses and Reviews.*

above B, but if B did not become extinct when it reached the Greenwich meridian, it would continue its eastward course, and C having in the meantime started on its first round, the fossils of the fauna B would be found both above and below those of C. It will be shown below that cases of recurrence do occur, but nowhere do we find a Silurian fauna above a Devonian one, or a Devonian one above one belonging to the Carboniferous, nor is the fauna of a great group of rocks found in one region above the fauna of another group, and in another region below the same. And this is true not only of the faunas of one major division, such as those of the Silurian and Carboniferous periods, but also of the faunas of many minor subdivisions into which the large ones are separated, for instance we do not find the Llandovery fauna of the Silurian period which in Britain is found below the Wenlock fauna embedded elsewhere in strata above the Wenlock. I have simplified the statement by assuming that the faunas are identical in the different localities, and exactly similar throughout the whole thickness of the containing strata, which is naturally not the case, but the additional complexity does not conceal the truth of what has been stated. In the absence of actual inversion of well-marked faunas, only one explanation is possible, namely, that the time for migration of forms is so short as compared with the entire period during which the forms existed, that it may be practically ignored, and the strata containing similar faunas may be therefore spoken of truthfully as contemporaneous and not merely homotaxial[1].

Apparent anomalies in the distribution of fossils. There are several occurrences which have tended to

[1] For fuller discussion of this matter see a paper by the Author 'On Homotaxis,' *Proc. Camb. Phil. Soc.*, vol. VI. Part II. p. 74.

augment the distrust frequently felt concerning the value
of fossils as indices of the age of the beds in which they
occur, which may be here considered.

Though the greater number of fossil remains belonged
to organisms which lived during the time of accumulation
of the deposits in which they are now embedded, this is
by no means universally the case, and the occurrence of
remanié fossils, which have been derived from deposits
more ancient than the ones in which they are now found
is far from being a rare event. The existence of remains
of this nature in the superficial drifts and river-gravels of
our own country has long been recognised, and no one
would suppose that the *Gryphœœ* and other shells
furnished by these gravels had lived contemporaneously
with the species of *Corbicula, Unio* and other molluscs
which are part of the true fauna of the gravels. In
this case the water-worn nature of the remains is a good
index to their origin, but in other cases, it is by no
means an infallible guide, for we sometimes find on the
one hand that remains of organisms proper to the deposits
in which they occur are water-worn, whilst on the other
the relics of *remanié* fossils are not. The now well-known
gault fossils of the Cambridge Greensand at the base of
the chalk were not always recognised as having been
derived from older beds, and there are certain fossils found
in nodules in the Cretaceous rocks of Lincolnshire, which
still form a subject for difference of opinion, for while
some writers maintain that they belong to the deposits
in which they are now found, others suppose that the
nodules have been washed out of earlier beds.

Occasionally we find forms which occurring in a set of
beds *A* in an area, are absent from the overlying beds *B*,
and appear again in the succeeding deposits *C*. Such cases

of *recurrence* are by no means rare, though many supposed
instances of recurrence have been recorded as the result
of- stratigraphical or palæontological errors. The best
examples have been noted by Barrande among the Lower
Palæozoic deposits of Bohemia. The stage D of Bohemia
consists of five 'bandes' or subdivisions, the lowest (d 1),
central (d 3) and uppermost (d 5) divisions are mainly
argillaceous, whilst the second (d 2) and fourth (d 4) are
essentially arenaceous. Some of the forms found in d 1,
d 3 and d 5 have not been found in d 2 and d 4. The
best-known example is the trilobite *Æglina rediviva*. It
is clear that this and other forms did not become extinct
during the deposition of the strata of d 2 and d 4, though
they may have disappeared temporarily from the Bohemian
area, or else lingered on in such diminished numbers that
their remains have not been discovered. The range of the
organism is in fact right through the deposits of the
stage D, and the discontinuity of distribution is not a real
anomaly; it may be compared to some extent with cases
of discontinuous distribution in space. It is needless to
remark that the whole fauna does not disappear for a time
and then reappear, but only a few out of the many forms
which compose it. The comparative rarity of examples of
recurrence after long intervals is an indication that the
palæontological record as it is termed is not so imperfect
as some suppose, for if our knowledge of fossils were very
imperfect, we should expect cases of apparent recurrence
to be common, as the result of the non-detection of fossils
in the intermediate beds. One of the most marked cases
of apparent recurrence known some years ago was the
reappearance of a genus of trilobite *Ampyx* in Ludlow
rocks, found in the Bala rocks, but not in the Llandovery
or Wenlock strata. It has since been discovered in

Llandovery beds, and its eventual discovery in beds of Wenlock age may be regarded as certain. A supposed case of recurrence which would have been remarkable, that of the disappearance of *Phillipsia* in Ordovician rocks, its entire absence in those of Silurian age, and its reappearance in the Devonian, has broken down, for the supposed Ordovician form has been shown to belong to an entirely different group of trilobites from that containing the genus *Phillipsia*, and it has been therefore renamed *Phillipsinella*.

Many apparent anomalies of distribution have been explained as due to migration, but it is doubtful whether any one of these supposed anomalies is actual and not due to errors in determining the position of the beds or the nature of their included fossils. Some of the supposed anomalies have already been shown to be due to error, and the others will almost certainly be cleared up. In speaking of anomalies of distribution, the geologist can only be guided by experience as to what constitutes an anomaly. For instance the existence of a complete fauna in any one place in the beds of a system above that to which it is elsewhere confined would be regarded as anomalous and as probably due to error, whilst the reappearance of several forms in beds of a system higher than that in which they had hitherto been found, could hardly be considered as an anomaly. A geologist would suspect the statement that after the disappearance of an Ordovician fauna in an area and its replacement by a Silurian fauna, the Ordovician fauna reappeared for a time, but would not regard the statement that a Cenomanian fauna partly reappeared in the Chalk Rock with surprise.

The existence of a Silurian fauna in Ordovician times

was maintained by Barrande in the case of the Bohemian basin. Lenticular patches of Silurian rocks having the lithological characters of the Silurian strata are found in the Ordovician beds of that region, and they contain fossils specifically identical with those of the Silurian rocks. Barrande explained this appearance as due to the existence of a fauna in other regions resembling the Silurian fauna of Bohemia, during the Ordovician period, when the normal Ordovician fauna of Bohemia inhabited that area. He supposed that in parts of the basin, when favourable conditions arose, *colonies* of the foreign fauna settled for a time, but did not get a permanent footing in the basin until the commencement of Silurian times. The theory of colonies has now been rejected for the Bohemian area, and the phenomena shown to be due to repetition of strata by folding and faulting, but it is a theory which is again and again advocated in order to explain apparently anomalous phenomena in other areas, and these apparent anomalies which are so explained, must be regarded with grave suspicion.

The various complexities alluded to in the foregoing pages increase the difficulty experienced by the geologist in correlating strata in different areas by their included organisms, but no one of them disproves the possibility of making these correlations, which can be carried on to a greater or less extent according to the nature of the faunas.

A good deal of misconception has arisen concerning the geographical distribution of former faunas, owing to the tendency to compare them exclusively with the littoral faunas of the present day. These littoral faunas have a comparatively limited geographical distribution, the forms of one marine province often differing considerably from those of an adjoining one, and still more widely from one

which is remote, so that anyone confronted with the relics
of faunas from the existing Australian and European seas,
would find no indications furnished by identity of species
that the faunas were contemporaneous. Recent researches
have shown, however, that the creatures whose remains
are deposited at some distance from the coastline have a
much stronger resemblance to one another than the
littoral organisms have, if the fauna of two distant
areas be compared. It is still a moot point which will
be discussed in a later chapter, how far the deep-sea
deposits of modern times are represented amongst the
strata of the geological column by deposits of similar origin.
But it is certain that many of the ancient strata are
not littoral deposits, and it will be found that it is by
comparison of the faunas of the deeper-water deposits
that the geologist correlates the strata of remote regions:
where shallow water deposits are formed, the faunas differ
markedly in different regions, and these shallow-water
forms can only be correlated owing to their occurrence
between deeper-water strata.   Thus if strata A, B and C
be found in one area, and the fauna of A and C are deep-
water forms, those of B being shallow-water forms, and
in another area beds A' contain the same fauna as A, and
C' the same fauna as C whilst the fauna of B' is different
from that of B, we can nevertheless correlate the strata
B and B' (if they be conformable with the underlying and
overlying beds), because of the identity of age of the
associated beds in the two areas. It will possibly be
found that the strata A and C can be further subdivided
into A_1, A_2,... &c. C_1, C_2,... by the existence of minor
faunas, which are comparable in the two cases, but such
subdivisions may not be established in the case of the
beds B and B'.

To take actual examples:—The Llandovery beds of Dumfriesshire can be subdivided into several minor divisions each of which can be recognised in the Lake District of England, and to a large extent in Scandinavia and elsewhere, for the deposits in these areas are of deep-water character, and the sub-faunas of the subdivisions are similar in the different areas, but the Llandovery rocks of the Welsh borderland are shallow-water deposits, with a different fauna from that of the deep-water deposits of this age, and can only be stated to be contemporaneous with the Llandovery rocks elsewhere, because the deeper-water faunas of the underlying Bala rocks and overlying Wenlock rocks of the Welsh borders are respectively similar to those of the Bala and Wenlock rocks of the other regions. The shallow-water Llandoveries of the Welsh borders have only been separated into two divisions, upper and lower, and have not been split up into a number of subdivisions, each characterised by a sub-fauna, and each comparable with one of the subdivisions of Dumfriesshire, Lakeland and the other regions where the deep-water acies is found.

It will be seen that though the principle of William Smith that strata can be recognised by their included organisms has been extended since his time, and shown to apply to far smaller subdivisions of the strata than was suspected, the method of application is the same, and is more or less successful according to the amount of evidence which is accumulated in support of it.

CHAPTER VI.

METHODS OF CLASSIFICATION OF THE STRATA.

EARTH-HISTORY like human history is the record of an unbroken chain of events. The agents which have produced geological phenomena have been in operation since the earth came into existence. Accordingly a perfect earth-history would be written as a continuous narrative, just as would a complete history of the human race. The historian of man finds it not only convenient but necessary to divide the epoch of which he is writing into periods of time, and so does the geologist, and in each case the division is necessarily more or less arbitrary. It is true that in writing the history or geology of a country, marked events stand out which form a convenient means of making divisions, but the marked events occurring in one country are not likely to take place simultaneously with those of another country, and consequently a classification of this character is only locally applicable.

The classification which is at present used by geologists was originally founded upon definite principles, and although our principles of classification have, as will appear, been somewhat altered subsequently, it has been found more convenient to modify the original classification than to adopt a new one in its entirety.

The largest divisions into which the strata of the geological column were separated were instituted because of the supposed extinction of faunas, and sudden or rapid replacement by other faunas of an entirely different character. This supposed rapid extinction and replacement is now known to have been only apparent and due to observation in restricted areas, and it is doubtful whether the three great divisions founded upon them are not rather mischievous than useful, as tending to disseminate wrong notions.

Moreover there is considerable diversity of opinion as to the terms to be adopted. The rocks were formerly divided into Primary, Secondary, and Tertiary. Owing chiefly to the use of the term Primary in another sense, the alternative titles Palæozoic, Mesozoic and Cainozoic (or Cænozoic) were suggested, and though the term Primary has been definitely abandoned in favour of Palæozoic, the words Secondary and Tertiary are used extensively as synonyms of Mesozoic and Cainozoic. It was soon perceived that the period of time included in the Palæozoic age was much longer than the combined periods of Secondary and Tertiary ages, and it was proposed to group the latter under one title Neozoic, whilst another suggestion was to split the Palæozoic age into an earlier Proterozoic and later Deuterozoic division. The interest excited by the advent of man is probably the cause of the attempt to establish a Quaternary division, which some hold to be a minor subdivision of the Tertiary, whilst others would separate it altogether. The terms Palæozoic, Mesozoic (or Secondary) and Cainozoic (or Tertiary) are now used so generally that any attempt to abolish them would be doomed to failure, but it must be remembered that they are purely arbitrary expressions, and the other terms which are not in general use, might be dropped with advantage.

The other subdivisions have been used somewhat loosely, and although an attempt has been made by the International Geological Congress to restrict certain names to subdivisions of varying degrees of value, it will probably be found best to allow of a certain elasticity in the use of terms, merely agreeing that they shall be used as nearly as possible with the signification assigned to them by the Congress. According to this classification, and apart from the division into Palæozoic, Mesozoic and Cainozoic, the strata of the geological column are grouped into *Systems*, which are subdivided into *Series*, and the series are further split up into *Stages*. A number of chronological terms were also suggested, of equivalent importance, thus the beds of a *system* would be deposited during a *Period*, those of a *series* during an *Epoch*, and those of a *stage* during an *Age*[1].

The rocks of the Geological Column were originally divided into systems, owing to the occurrence of marked physical and palæontological breaks between the rocks of two adjacent systems, except in cases where a complete change occurred locally in the lithological characters of the rocks of two systems which were in juxtaposition: it is necessary to consider for awhile the nature of these breaks.

The most apparent physical break is where the rocks of one set of deposits rest unconformably upon the rocks of another one, indicating that the older set has been

---

[1] The chronological words have been used so loosely that it is doubtful whether any good will come of trying to restrict their use, and Sir A. Geikie has pointed out the confusion which would arise if the term *group* be employed for the largest divisions (Palæozoic, &c.). The terms *System*, *Series* and *Stage* may well be employed in the senses suggested by the Congress.

uplifted and to some extent eroded before the deposition of the strata of the newer set. This uplift and erosion signifies a change from oceanic to continental conditions in the area in which unconformity is found on a large scale, and accordingly a long period of time would elapse during which the continental surface would not receive deposits, so that the highest rocks of the underlying system would be considerably older than the lowest rocks of the one which succeeds it. Such a break may be obviously utilised for purposes of classification, but as some areas of the earth's surface must have been occupied by the waters of the ocean when other regions formed land, deposit in some areas must constantly have occurred simultaneously with denudation in others, and any classification founded upon the existence of unconformities will therefore have a purely local value.

Another, and less apparent physical break, which will also be locally applicable, may be due to the depression of an area to so great a depth that little or no deposit was formed upon the ocean floor there during the period of great depression; but as a break of this character is difficult to detect, the existence of unconformities has alone been practically utilised as a means of separating strata into systems owing to marked physical change, except in the cases where the lithological character of the strata completely changes, as between the Triassic and Jurassic rocks of England.

Palæontological breaks or breaks in the succession of organisms are in many cases, the result of physical breaks, and accordingly it is often possible to separate one set of strata from another by the existence of a combined physical and palæontological break between them. It is by no means necessary however that a physical break should be accompanied by a break in succession of the organisms, and the latter may also occur without the former. It was once maintained that a palæontological break was due to the complete and sudden extinction of a fauna and its entire replacement by a new one, but this is far from true, and accordingly the breaks differ in degree. Study of the strata shows that when the succession is not to any extent interrupted, the species do not appear

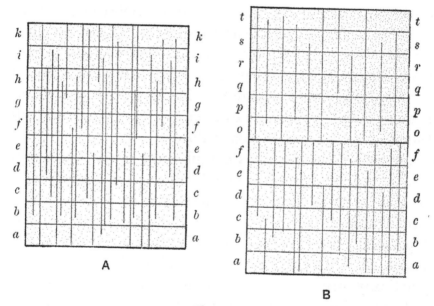

Fig. 2.

simultaneously, but come in at different horizons, and they disappear in the same way. In Figure 2 let *A* represent a set of conformable strata *ab...k*, and suppose the vertical lines represent the ranges of the various species found in these strata. It will be seen that of 27 species whose range is shown only 2 pass through the whole thickness, so that the fauna of *k* is very different from the fauna of *a*, nevertheless the fauna of each stratum is closely similar to that of the underlying as well as to that of the overlying stratum, and though most of the species of k are different from those of a, this need not be the case with the genera. The fauna of the set of strata would contain every species whose range is represented, and for convenience' sake it might be said to be composed of sub-faunas, one of which occurs in each division $ab...$, but the separation into sub-faunas would be artificial and merely for convenience' sake, for there is no break between any two sub-faunas. Turning now to B (Fig. 2), an attempt is made there to show what happens when there has been a physical break, resulting in the denudation of the strata $ghik$, and the deposition of another set $op...$ unconformably upon those deposits of the earlier set which have not been denuded. As the result of this we note, first, that the relics of organisms which existed in the area during the deposition of $ghik$, and were entombed in those strata, are destroyed by the processes of denudation, and a large number of organisms which lived long after the deposition of f, and disappeared not simultaneously but at different times during the period when denudation was in operation, seem to become extinct simultaneously at the top of f, though, if we could visit an area which was receiving sediment during the period of denudation, we should find them dying out in the rocks of that region at different levels. Furthermore, whilst denudation is going on, a longer or shorter period of time elapses, during which the upheaved area receives no deposit, and accordingly no organisms which lived during that period are preserved in the upheaved area. During this time a set of deposits lmn may have been laid down elsewhere, and besides the gradual disappearance of some of the organisms of $ab...k$, there will have been a gradual appearance of new species. When the upheaved area is once more submerged, a new set of deposits *op*... is accumulated in it, and the species which gradually appeared in adjoining regions will now migrate to it, and will seem to come in simultaneously at the bottom of *o*; accordingly we may find that there is not a single species which passes through from *f* to *o* and the palæontological break in this area is complete, though it is clear that it only implies local change, and that we may and indeed must find intermediate forms in other regions which fill up the gap.

As an illustration of the local character of a palæontological break we may cite the case of the Carboniferous and Permian systems of Britain. These rocks are separated from one another in our area by a physical and palæontological break, but in parts of India, and other places, we find a group of rocks now known as the Fermo-Carboniferous rocks which contain a fauna intermediate in character between those of the Permian and Carboniferons systems, and a study of this fauna shows that the hiatus which exists locally is filled by the species contained in the Fermo-Carboniferous rocks.

A palæontological break may, like a physical one, result from depression of the ocean-floor to so great a depth, that no organisms are preserved there during the period of great depression, and the remarks made concerning a depression of this nature when speaking of physical breaks will apply here also.

A local palæontological break may result owing to physical changes without the production of an unconformity in the area, or its submergence to a great depth, or if an unconformity is found, the break may be more marked owing to other physical changes. The difference between the Upper and Lower Carboniferous faunas is very marked in England, where the Upper Carboniferous beds were deposited under physical conditions different from those of the Lower Carboniferous, and accordingly the corals, crinoids and other open-water animals which flourished in Lower Carboniferous times are rare or altogether absent in the higher rocks. Where the change of conditions did not occur to a great extent as in parts of Spain and North America, the similarity between the two faunas is much more pronounced. Again, there is an unconformity between the Cretaceous and Eocene beds of England, which is accompanied by a palæontological break, but this break is more pronounced owing to difference of physical conditions, for we find abundance of gastropods in the lower Tertiary beds, and a rarity of these shells at the top of the chalk of England, though where physical conditions were favourable for the growth of gastropods, their shells are found in the higher strata of chalk age, and the palæontological break is not so apparent.

A palæontological break may occur also as the result of climatic change, though actual instances of this occurrence are much more difficult to detect owing to the general absence of any evidence of climatic change other than that supplied by the organisms themselves. Still, when no physical break exists, and the lithological characters of a group of sediments remain constant throughout, indicating the prevalence of similar physical conditions through the period of deposition of the sediments, if the fauna suddenly changes, there must have been cause for the change, and in the absence of any other cause which is likely to produce the change, alteration of the character of the climate may be suspected.

It follows from the observations which have been made, that although the rocks of the Geological Column may be divided into systems owing to the existence of physical and palæontological breaks, and this classification may be and has been applied generally, the line of demarcation between the rocks of two systems will be a purely conventional one, where there is no break, and, to avoid confusion, that line when once drawn should be adopted by everyone, unless good cause can be shown for its abandonment.

The subdivision of systems into series has been conducted in a manner generally similar to that in which large masses of strata have been grouped into systems, with the exception that actual breaks need not occur. The subdivision was usually made on account of marked differences in the lithological characters or fossil contents of the rocks of the various series, and frequently the lithological characters as well as the fossil contents are dissimilar ; taking the rocks of the Silurian system of the typical Silurian area as an example, we find the Llandovery rocks largely arenaceous, the Wenlock rocks largely calcareo-argillaceous, and the Ludlow rocks argillaceo-arenaceous, whilst the fauna of the Wenlock rocks differs from that of the Llandovery rocks below and also from that of the Ludlow rocks above. The Llandovery, Wenlock and Ludlow therefore constitute three series of the Silurian system, but the lines of demarcation between these series are nevertheless conventional, for it has been suggested that a more natural division, as far as the British rocks are concerned, could be made by drawing a line, not as at present at the base of the Ludlow, but in the middle of that series as now defined, and uniting the Lower Ludlow beds with the Wenlock strata to form a single series.

The same process as that adopted in the case of series has been essentially pursued in subdividing these into stages. Each stage is usually different from that above and below in its lithological characters, fossil contents, or both, though the difference is usually less in degree than that which has been utilised for the demarcation of series. A stage is often, though not always, composed of deposits of one kind of sediment, and is furthermore frequently characterised by the possession of one or, it may be, two, three or more characteristic fossils. Thus the Wenlock series is divided in the typical area into Woolhope lime-stone, Wenlock shale, and Wenlock limestone, and the very names given to these stages indicate that each is largely composed of one kind of material. Their fossils are also to some extent different, though the difference between them is not likely to be of so marked a nature as that which exists between the faunas of separate series.

It will be seen that the system differs from the series and the series from the stage in degree rather than in kind, and no hard line can be drawn between divisions of different degrees of magnitude. It follows therefore that frequently a mass of sediment which one author will consider sufficiently important to constitute a system will be defined by another as a series, and similarly a series of one writer may become a stage of another.

The student of Stratigraphical Geology will find the expression 'fossil zone' occurring over and over again in geological literature, and as the term has been used some-what vaguely by many writers and is apt to be misunder-stood, it will be useful to notice the expression at some length.

Strictly speaking the term zone (a belt or girdle), when applied to distribution of fossils, should refer to the belt of strata through which a fossil or group of fossils ranges. Generally speaking, the expression is used in connexion with one fossil; thus we speak of the zone of *Cænograptus gracilis*, the zone of *Cidaris florigemma* and the zone of *Belemnites jaculum*, though sometimes it is used with reference to more than one species, as the zone of Micrasters and the *Olenellus* zone. The term has been used not of a belt of strata but of a group of organisms[1], and zones defined as " assemblages of organic remains of which one abundant and characteristic form is chosen as an index," but if it be agreed that the term should be applied to strata and not to organisms this might be modified and the definition run :—' Zones are belts of strata, each of which is characterised by an assemblage of organic remains of which one abundant and characteristic form is chosen as an index.'

It has been objected that the subdivision of strata into zones has been pushed too far, but this is merely because in the establishment of zones, workers find it easier to work out the successive zones where the strata are thin and presumably deposited with extreme slowness, than where they are much thicker and have been rapidly accumulated, and accordingly, as the subdivision of strata into zones is a recent event, geological literature contains many more references to thin zones than to those of great thickness. Where an abundant and characteristic form (which is chosen as an index) of an assemblage of organic remains ranges through a great thickness of deposit, there is no objection to speaking of the whole as a zone, and it cannot be divided. To give some idea of the variations in the thickness of strata through which these abundant

[1] See H. B. Woodward, "On Geological Zones," *Proc. Geol. Assoc.*, vol. XII. Part 7, p. 295, and vol. XII. Part 8, p. 313.

and characteristic forms will range, I append a list of the
zones of graptolites which have been established amongst
the Silurian rocks of English Lakeland and the thickness
of each (which in the case of the thicker deposits is
naturally only approximate) :—

Zone of	Thickness.	
	Feet.	Inches.
Monograptus leintwardinensis	5000	0
Monograptus bohemicus	5000	0
Monograptus Nilssoni	1000	0
Cyrtograptus Murchisoni	1000	0
Monograptus crispus	22	0
Monograptus turriculatus	60	0
Rastutes maximus	25	0
Monograptus spinigerus.....................	3	0
Monograptus Clingani	3	0
Monograptus convolutus.....................	7	6
Monograptus argenteus	0	8
Monograptus fimbriatus	7	6
Dimorphograptus confertus	25	0
Diplograptus acuminatus	2	6

It must not be supposed that each of the subdivisions
in the above list is of equal importance, and has occupied
approximately the same length of time for its formation,
but a study of the strata proves by various kinds of
evidence that the deposits in which the characteristic
forms range through a small thickness of rock were on the
whole deposited much more slowly than where the range
is continuous through a great thickness of deposit.

The geological systems, as originally founded, were
not very accurately separated from one another except
locally. A comprehensive view of the characters of a
system was taken, and accordingly the lines of demarcation between the same systems adopted by workers in

different countries were by no means necessarily at or near the same geological horizon. As the result of more recent work, the establishment of fossil zones has been growing apace, and though many of these are seen to have only local significance, it is found as the result of experience that many of them are widely spread and occur in the same order in different localities; accordingly the remarks that have been made concerning the contemporaneity of strata apply to these zones also. After a study of this kind, a much more accurate comparison of strata is possible, and correlation of strata can be carried on to a much greater extent than when the systems were only roughly subdivided by reference to breaks, differences of lithological character, and general comparison of the faunas; accordingly whilst largely retaining the old names, the old method of classification is being partly superseded, and the included faunas alone are utilised to establish accurate correlations of the strata in various parts of the world. How far this correlation can be carried on remains to be seen, for the work though well advanced has by no means reached completion, and predictions as to the ultimate issue are useless without the experience by means of which only the work can be done. The difference between the methods of classification is well shown by an examination of the old and new divisions of the chalk. It was formerly roughly divided mainly by lithological characters into Chalk Marl, Lower Chalk without flints, Middle Chalk with few flints and Upper Chalk with many flints, but no two observers would probably agree as to where the deposit with few flints ceased and that with many commenced. The chalk is now separated on palæontological grounds into Cenomanian, Turonian, Senonian and Danian, and the superiority of the new method to the old is practically shown by the abandonment of the old classification except for very rough purposes, and the general acceptance of the new one. Many other examples might be given, but this one will suffice. In the case of some of the systems, the Carboniferous for example, the old classification founded upon lithological characters is largely extant, and it has been inferred therefore that no accurate subdivisions of the Carboniferous rocks can be made by reference to the faunas, owing to the rapidity with which the deposits were accumulated. It is by no means certain because the work has not been done that it cannot be done, and the experience obtained from a study of other strata in which subdivisions have been established by reference to the fauna would lead one to suppose that the non-establishment of subdivisions of the Carboniferous strata is due to our want of knowledge rather than to their non-existence.

The establishment of a classification on palæontological lines by no means does away with the necessity for local classifications on a lithological basis, and it has already been remarked that important results will follow from a comparison of the classifications of sediments founded on the two lines, results which have hitherto largely escaped our attention owing to the existence of a cumbrous classification attained by the application sometimes of one method, at other times of the alternative one.

CHAPTER VII.

SIMULATION OF STRUCTURES.

ALTHOUGH it is easy to give an account of the structures which are of importance to the student of the stratified rocks, actual observation of these structures is frequently attended with difficulties owing to the close imitation of one structure by another, and the past history of the science shows that erroneous conclusions have been reached again and again on account of the incorrect interpretation of structures.

Simulation of organisms has frequently been the cause of error. Inorganic substances take on the form of organisms with various degrees of closeness. The dendritic markings produced by efflorescences of oxide of manganese are familiar to all, and as the name implies, they simulate, to some extent, plant remains. More complex chemical changes have resulted in the production of rock-masses in which, not the outward form alone but, the internal structure of organisms is reproduced with more or less approach to fidelity, as the rocks which contain the supposed organisms described as *Eozoon bohemicum, E. bavaricum,* and, we may add, *E. canadense.* Mechanical changes in rocks subsequent to their formation may also cause the simulation of organisms by inorganic substances. Prof. Sollas has given reasons for considering the structure described as *Oldhamia* to be inorganic, and in the Carboniferous Sandstones of Little Haven, Pembrokeshire, every stage in the formation of tubular bodies resembling worm-tubes, as the result of complex folding of the strata, may be observed, whilst in other cases we find imitation of worm-tracks, as has been observed before.

It is when one inorganic structure is simulated by another that the stratigraphical geologist is most likely to be led astray, and accordingly it is worth noting some cases where this has occurred, as a warning, for it must not be supposed that the cases here noted are the only ones which are likely to occur.

It has been seen that the existence of bedding-planes is of prime importance to the geologist, and their detection is a matter of supreme moment. Under ordinary circumstances there is no great difficulty in distinguishing bedding-planes from other planes, but the importance of discovering them is often greatest when the difficulty is most pronounced. In rocks which have undergone no great amount of disturbance the planes of stratification are often marked by their regular parallelism, the separation of layers having different lithological characters by these planes, the arrangement of the longer axes of pebbles parallel to them, and the occurrence of fossils and also of rain-prints, ripple-marks and other structures produced during deposition, upon the surfaces of the strata, but none of these appearances is necessarily conclusive, especially in areas where the rocks have been subjected to orogenic movements. In regularly-jointed rocks, jointing may well be mistaken for bedding, and there is often great difficulty in discriminating between bedding and cleavage, especially when the exposures of rock are of small extent. Fossils may be dragged out along planes

at an angle to the true bedding, pebbles will be compressed by cleavage so that their longer axes do not
remain parallel to the bedding-planes but now lie parallel
to the superinduced planes of cleavage, and a structure
closely resembling 'ripple-mark' may be produced on
planes other than those of original bedding, as the result
of puckering. The alternation of rocks having different
lithological characters may also be misleading. Intrusion
of dykes along cleavage-planes, followed by decomposition
of the dyke-rock causing it to resemble a sediment, and
formation of mineral veins along the same planes, may
give rise to an apparent succession of rocks of different

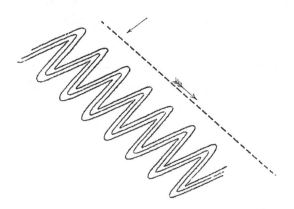

FIG. 3.

lithological characters which could easily mislead an
observer and cause him to mistake the cleavage-planes
for planes of stratification. In rocks which have undergone great lateral pressure, the beds of different lithological character may be folded in such a way as to give
very erroneous ideas of the true dip of the rock on a
large scale. In Fig. 3 the dip of the rocks in a small
exposure might appear to be in the direction indicated by
the unfeathered arrow, whilst the true dip of the strata as

a whole, leaving the minor foldings out of account, is in the direction of the feathered arrow, at the inclination represented by the dotted line. The minor folds in a case like that represented may extend upwards for scores or even hundreds of feet, so that an error as to the direction and amount of dip may be made, even if the observer faces a cliff of considerable height.

False-bedding on a large scale may be a cause of error. In the Penrith Sandstone of Cumberland, the planes of deposition are often found dipping in one direction in a large quarry, but inspection of a wider area shows that this is not the true dip of the beds as a whole, but merely a local dip due to deposition on a slope, and any one attempting to calculate the total thickness of the beds by reference to these divisional planes might be seriously led astray. A reference to Fig. 4 will explain this. The lines *AA'*, *BB'* are the true bedding-planes

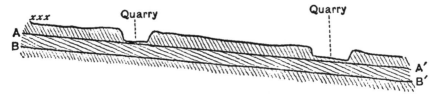

FIG. 4.

cut across in the section, whilst the lines sloping to the right from *xx* are only lines of false-bedding on a large scale. An exaggerated estimate of the thickness of the deposit would be made by measuring the thickness of each of these stratula from *A* to *A'* and adding these thicknesses together, whereas the actual thickness of the middle bed is the distance between *A* and *B* or *A'* and *B'*.

When rocks have been affected by thrust-planes, the simulation of bedding may be carried out to a very full extent. Not only do the major thrust-planes resemble bedding-planes but the minor thrusts produce an appearance of divisional planes separating stratula or laminæ, and a close approximation to false-bedding is the result. To this structure Prof. Bonney has given the name 'pseudo-stromatism[1].' It may be developed in rocks of all kinds, whether possessing original planes of stratification or not, and as a result of its existence the geologist may be seriously misled, not merely by mistaking the direction of the strata, but also the nature of the rock, for we may find it produced in an unstratified glacial till, and in a massive igneous rock, and in each case the resulting rock will resemble a sedimentary deposit, and of course the observer may be confirmed in his erroneous opinion by the formation of apparent fossils, ripple-marks or other objects which he might expect to discover in sediments. As illustrative examples, reference may be made to a number of schistose rocks, in which the planes of dis-continuity (which are in truth planes of foliation) have been taken for bedding-planes and the rocks claimed as sedimentary though they are in reality igneous; for instance many of the rocks of the Laurentian of Canada, of the Hebridean of the North West Highlands, and some of the ancient rocks of Anglesey.

A foliated structure may, as is now well known, be simulated by a structure developed in a rock prior to its consolidation. The similarity of flow structure of some lavas to the foliated structure of a schist was long ago pointed out by Darwin and Scrope, and recent work has proved that parallel structure due to differential move-ment prior to consolidation may be developed in plutonic

---

[1] Bonney, T. G., *Quart. Journ. Geol. Soc.*, vol. XLII. *Proc.* p. 65.

rocks, as shown by Lieut.-General McMahon in the Himalayan granites, and by Lawson amongst the plutonie rocks of the Rainy Lake Region; and as the foliated structure may be mistaken for original stratification the same may occur, and has occurred, when dealing with this flow-structure.

This is not the place to discuss the truth of the old theory of progressive metamorphism, in which it was maintained that a gradual passage could be traced between ordinary sediments and plutonic rocks, but it may be pointed out that much of the evidence which was relied upon to prove the theory was fallacious and due to the confusion of the parallel structure set up in plutonic rocks prior to, or subsequent to, consolidation, with original stratification. Recent study of metamorphic rocks has proved that the parallel structures developed in the rocks of an area which has undergone metamorphism may be produced by three distinct processes; they may be original planes of deposition, or formed in a solid rock subsequently to its formation, or in an igneous rock before its consolidation, and although it is sometimes possible to separate the structures produced by these processes, this is not always the case[1]. When a plutonie rock contains large phenocrysts and an eye-structure is developed in it, it may simulate a conglomerate, the rounded phenocrysts being taken for pebbles[2]. Still closer simulation of an

[1] It must be noticed that the rock in which parallel structure is produced before consolidation, if it undergoes no further change, though often associated with metamorphic rocks, is not itself metamorphic. The term *gneiss* applied to these rocks is a misnomer, unless the term be used even more vaguely than it is at present.

[2] See Lehmann, *Untersuchungen über die Entstehung der Altkrystallinischen Schiefergesteine mit besonderer Bezugnahme auf das Sächsische Granulitgebirge*, Plate XI. fig. 1.

epiclastic conglomerate may be produced in other ways
and will be referred to immediately.

We have already seen that the existence of unconformities has been utilised in the demarcation of large
divisions of strata in various regions, and whether they be
utilised in this manner or not, their detection is a matter
of importance to the stratigraphical geologist, as they
afford information concerning the occurrence of great
physical changes during their production. These unconformities may also be closely simulated by structures
produced in very different manner.

The occurrence of an unconformity implies the denudation of one set of beds before the deposition of another
set upon them, and accordingly the denuded edges of the
lower set will somewhere abut against the lower surface
of the lowest deposit or deposits of the overlying set[1].
The existence of an unconformity may often be detected in
section, but when the unconformity is upon a large scale
this may not be possible, but it will be discovered by
mapping the strata and will be apparent on a map owing
to the deposits of the lower set of beds abutting against the
others. This is well seen where the Permian rocks of
Durham, Yorkshire, and Nottinghamshire rest upon different members of the underlying Carboniferous series, and

[1] An unconformity may be simulated or an actual unconformity
rendered apparently more important, as the result of underground
solution of the underlying strata subsequently to the deposition of the
upper set upon them, and any insoluble materials in the underlying
strata may be left as an apparent pebble-bed at the base of the upper
beds. This is seen at the junction of the Tertiary beds with the chalk
near London. Subterranean water has dissolved the upper part of
the chalk, increasing the unconformity which naturally exists between
chalk and Tertiary beds, and the insoluble flint of the dissolved chalk
is left as a layer of 'green-coated flint' at the base of the Tertiary
deposits.

will be noticed on any good geological map of England. But a similar effect may be caused by a fault, so that mere inspection of a map or even of the strata in the field and discovery of one set of beds ending off against another does not prove unconformity. When the fault is a normal one, with low hade (that is, having a fissure approaching the vertical position), the outcrop of the fault-fissure will approximate to a straight line if the fault has a straight course, even if the ground be very uneven, whereas, if the plane of unconformity has not been tilted to a high angle from its original horizontal position, it will crop out in a sinuous manner across uneven ground, in a way similar to that of beds which are nearly horizontal, so that though the general trend of the outcrop of the plane of uncon-formity may be fairly straight, its deviation from a straight line will be frequent and marked, as seen in the case of the Permian unconformity above referred to. But if the unconformable junction has been highly inclined its outcrop will resemble that of a normal fault, or if the fault be a thrust-plane with high hade, the outcrop of this will resemble that of an unconformable junction which has not been greatly tilted from its original horizontal position. In these cases we require more evidence before we can decide whether we are dealing with an unconformable junction or a faulted one.

The lowest deposits of the newer set of strata lying above an unconformity have probably been laid down in water near the shore-line. As the unconformity, if large, implies elevation above the sea-level, the deposits first formed after this elevation has ceased, and depression commenced, will necessarily be littoral in character and possibly of beach-formation, and accordingly we often find that an unconformity is marked by the existence of an epiclastic conglomerate immediately above the plane of unconformity and, although this need not be continuous, it is usually found somewhere along the line of junction. The conglomeratic base of the Lowest Carboniferous strata when they repose upon the upturned edges of the Lower Palæozoic rocks of the dales of West Yorkshire is well known, and may be cited as an example. The association of conglomerates with unconformities is indeed so frequent that its possible occurrence will always be suspected and sought by the geologist. Unfortunately the result of recent observation is to show that along thrust-planes of which the outcrop simulates those of unconformable junctions, the difficulty of discrimination may be increased by the existence of cataclastic rocks which bear a close resemblance to epiclastic conglomerates, and which may be and have been styled conglomerates. It is well known that fragments of the adjoining rocks are knocked into a fault-fissure during the occurrence of the movements which cause the fault, to constitute a *fault-breccia,* and as the result of the abrasion of these fragments by chemical or mechanical agency, the angular fragments may become rounded and converted into rounded pebble-like bodies, when the rock is changed into a *fault-conglomerate.* Fig. 5, from a photograph kindly supplied by Prof. W. W. Watts, shows a stage in the formation of a conglomerate of this nature from a fault-breccia; the fragment on the right remains angular, whilst those on the left have become much more rounded. The illustration is from a case described by Mr Lamplugh occurring in the slaty rocks of the Isle of Man, and Mr Lamplugh's paper[1] furnishes the reader with

[1] Lamplugh, G. W., "On the Crush-Conglomerates of the Isle of Man," *Quart. Journ. Geol. Soc.*, vol. LI. p. 563.

references to other examples of the production of similar rocks. No general rule can be laid down for distinguishing the true from the apparent unconformity, for the attendant phenomena will differ in each case; but if a fault-conglomerate should be suspected, the observer

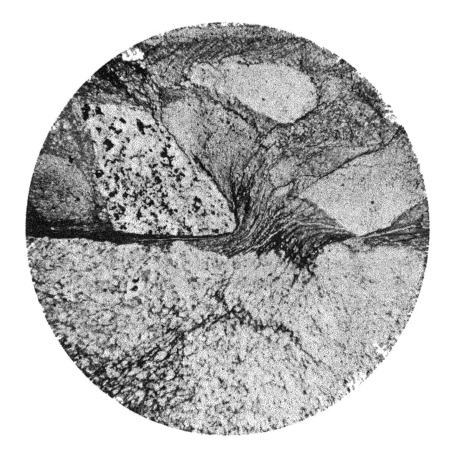

Fig. 5.

should try to ascertain whether fragments of a newer rock are imbedded in an older one, which sometimes occurs; he should note the existence of extensive slickensiding along the plane of junction and along planes of faulting, though the existence of these, implying as it does the

M. 6

occurrence of differential movement along the plane, does not prove that the movement was necessarily great, or that it did not take place along a plane of original unconformity; above all, he should look for structures such as mylonitic structure, pseudo-stromatism, development of new minerals, crushing out and stretching of fossils and fragments and, in short, for any structure which is familiar to him as a result of orogenie movements.

The effects of thrusting not only give rise to appearances suggestive of unconformity, but naturally also to a simulation of overlap. The thrust-planes are often parallel to original bedding-planes for some distance, but must cut across them sooner or later, producing lenticular masses which might be supposed to be due to the thinning out of beds as the result of cessation of deposition in a lateral direction.

Attention has already been directed to the deceptive appearance of great thickness of strata which is due to repetition of one stratum or set of strata by a series of thrust-planes, so that there is no actual inversion of any part of a bed. When masses of limestone are affected in this way, the thrust-planes may become sealed up, as the result of chemical change, and a compact irregular mass of limestone devoid of any definite divisional planes may be the consequence, and beds of grit sometimes exhibit the same feature to some extent.

Enough has been said to show that simulation of one structure by another has frequently occurred in rocks in so marked a degree as to render mistakes easy; and that these examples of 'mimicry' in the inorganic world are particularly frequent in rocks which have been subjected to great orogenic movements. The student will do well to acquaint himself with the macroscopic and microscopic structures which may be taken as characteristic of the rocks which have been thus affected, some of which can usually be detected with ease, and when he discovers them he may suspect that many phenomena which appear explicable in one way were in reality produced in a different one, for it is frequently very true of a region in which the rocks have been violently squeezed, stretched and broken that 'things are not what they seem.'

CHAPTER VIII.

GEOLOGICAL MAPS AND SECTIONS.

THE writer does not propose to give an account of the intricacies of geological mapping, for their right consideration requires a separate treatise[1]; all he desires is to call attention to some of the uses of geological maps as a means of conveying information. A geological map may be looked upon as an attempt to express as far as possible in two dimensions phenomena which possess three dimensions; this can be done to some extent on the actual surface of the map, by conventional signs, still more fully, by supplementing the map with sections; but best of all by a geological model, which is cut across in various directions in order to show the underground structure as well as that of the surface.

The ordinary geological map is one which shows the outcrop of the strata, subdivided according to age, as they would be seen upon the surface of the earth after stripping off the superficial accumulations, and it is to be feared that the term 'geological map' is associated in the minds of most students with a map of this character and of no other.

---

[1] The student is recommended to consult in particular, Appendix I. "On Geological Surveying" in *The Student's Manual of Geology*, by J. B. Jukes (Third Edition, Edited by A. Geikie), p. 747, and *Outlines of Field Geology*, by Sir A. Geikie (Macmillan and Co.).

Nevertheless, a great many most important observations other than those connected with the order of succession of the strata are capable of representation upon a geological map, and the possession of a large number of maps of any area upon the geology of which a person is engaged— each map to be used for recording observations of a particular kind—will save much writing in note-books and, what is of more importance, will allow him to compare observations which have been made at different times at a glance, instead of causing him to search through a series of note-books. Still, however well furnished with maps, the geologist will find a note-book essential[1].

The earliest geological maps represented the variations in the surface soils, or at most the general lithological characters of the rocks which by their decay furnished the materials for the soils. We have seen that the first chronological map was due to William Smith, and most subsequent English geological maps have been based upon his map of the strata of England and Wales. The order of succession of the strata is represented in these maps to some extent by the use of arrows to indicate the direction of dip of the strata, though this is not an unerring guide where strata are reversed, and accordingly the addition of a legend at the side of the map may be looked upon as essential to the correct understanding of the map itself. The legend is usually in the form of a section of a column, the strata being arranged in right order, the oldest at the base and the newest at the

---

[1] As a result of some experience, the writer recommends every student to acquire some skill in the use of the pencil, and if to such a degree that he can combine artistic effect with accuracy, so much the better. An acquaintance with photography is invaluable: often the possession of a camera would enable a section to be recorded, which is otherwise lost to science.

summit, the colours by which the strata are indicated being similar to those placed upon the map. Other information besides the mere order of succession of the strata may appear in the legend; thus their relative and actual thicknesses can be indicated if the column is drawn to some definite scale, and a brief description of the lithological characters of the rocks may well be appended to the side of the column. On the actual maps it is customary to exhibit the outcrop of the junctions of all igneous rocks as well as of the sedimentary ones: the nature of the metamorphism which sedimentary rocks have undergone at the contact with igneous ones may be and often is indicated by suitable signs; the position of faults is shown, and often also that of metalliferous veins, the nature of the ore in the latter being further indicated in some suitable manner, as by giving the recognised symbol for the metal; and in many maps an attempt is made to show the variations in dip and strike of the cleavage-planes.

The Geological Survey of the United Kingdom publishes two sets of maps, one showing the 'solid geology' and the other the 'superficial geology.' It is easier to understand these terms than to define them, for in Britain there is a sharp line between the two everywhere except near Cromer. The maps showing the superficial geology represent gravels, glacial drifts and other incoherent accumulations of geologically recent origin, which to a greater or less extent mask the strata below which are usually composed of more or less solidified material. The maps showing the solid geology display the outcrops of these strata, though it is usual to insert alluvium upon these maps, as it is often impossible to trace the junction-lines of the strata below it. Attention has already been directed to the fact that these maps of solid geology, though chronological, that is, having the strata represented according to age, are founded largely upon lithological differences, rather than upon included organisms; and it has been stated that for theoretical purposes two sets of chronological maps, · one founded upon lithological differences, the other upon difference of fossil organisms, would be extremely valuable.

Other phenomena are often best represented upon separate maps, for if all observations are crowded upon one map the result will be very confusing. Special glacial maps showing the contour of the country, with the portions between the contour lines coloured differently according to altitude, say the country between sea-level and 500 feet light green, that between 500 and 1000 dark green, that between 1000 and 1500 light brown and so on, exhibiting the direction of all observed glacial striae, the distribution of boulders so far as it is possible, and any other glacial phenomena which can be noted upon the map, will be valuable to the student of glaciation[1].

Various structural features may be well displayed on separate maps. The trend of the axes of folds will be useful, and may be accompanied by other information of cognate character[2]; maps of the distribution of joint

---

[1] For examples see Tiddeman, R. H., "Evidence for the Ice-Sheet in North Lancashire and the adjacent parts of Yorkshire and Westmorland," *Quart. Journ. Geol. Soc.*, vol. xxviii. pl. xxx., and Goodchild, J. G., "Glacial Phenomena of the Eden Valley" &c., *Quart. Journ. Geol. Soc.*, vol. xxxi. pl. ii.; and for a map of distribution of boulders, Ward, J. C., "Geology of the Northern Part of the English Lake District" (*Mem. Geol. Survey*), pl. iv.

[2] See Bertrand, M., "Sur le Raccordement des Bassins Houillers du nord de la France et du sud de l'Angleterre," *Annales des Mines*, Jan. 1893, Plate 1.

planes may be given in combination with those showing
the folding of the strata if it be desired to exhibit the
relationship between these; or with the physical features
of the country, if the dependence of physical features
upon joint structure be under consideration[1].   Much
information concerning cleavage may be acquired from a
map showing anticlinal and synclinal axes of cleavage[2], or
the actual strike of the cleavage over different parts of a
map may be represented, and its relationship to the
geological structure of the district exhibited[3].

Maps exhibiting changes in physical geography apper-
tain to the geologist as well as to the geographer.   The
position of ancient beaches, former lakes, representation
of the changes in the courses of rivers and kindred
phenomena may be shown upon maps, and will prove
useful[4].

A perusal of the maps to which reference has been
made above will give the student some notion of the
extent to which maps may be utilised to represent geological structures, and may suggest other methods by
which they may be utilised.

A geological section is usually drawn in order to
exhibit the lie of the rocks, as it would be seen if a
vertical cutting were made in that part of the earth's

[1] See Daubrée, A., *Études Synthétiques de Géologie Expérimentale*,
1ère Partie, Plates III.—VI., for an example of the latter, which is also
interesting as showing the utility of a map on transparent paper super-
posed on another, when illustrating the connexion between two sets
of structures.

[2] Ward, J. C., *Geology of the Northern Part of the English Lake
District*, Plate IX.

[3] Harker, Alfred, "The Bala Volcanic Series of Caernarvonshire"
(*Sedgwick Essay* for 1888), Fig. 5.

[4] For examples of maps of this kind, see Kjerulf, Th., *Die Geologie
des südlichen und mittleren Norwegen.*

crust which is under consideration. The character of the
section will depend upon circumstances. The Geological
Survey of Great Britain issues two kinds of sections which
are usually spoken of as vertical sections and horizontal
sections, though each is in truth a vertical section; but
whereas in the former the horizontal distance represented
is small as compared with the thickness of the strata, in
the latter the rocks of a considerable horizontal extent of
country are exhibited in the section, and the section is
not carried down to a great depth below the earth's
surface. There is no essential difference between the
two kinds of section, and often sections are drawn which
cannot be definitely classed as belonging to either kind,
but in extreme cases the vertical section is a representation of the order of succession as it would appear if the
rocks were horizontal, no matter how disturbed they may
be in reality; whereas the horizontal section represents
the strata as they actually occur, with all the folds and
faults by which they are affected. The accompanying
figure (Fig. 6) represents a horizontal section on the left
side of the figure with a vertical section of the same rocks
on the right side.

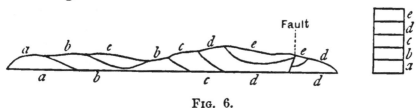

Fig. 6.

Vertical sections are extremely useful when it is
desirable to compare variations in the strata over wide
extents of country: this can be done by drawing a series
of columns of the strata, each showing in vertical section
the lithological characters and thicknesses of the strata in

one place, whilst the relationship between the strata of
two different places may be indicated by joining the beds
of the same age by dotted lines as shown in Fig. 7[1].

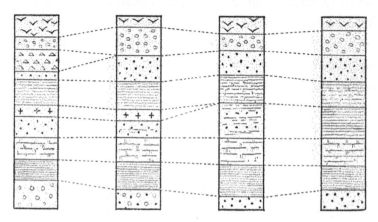

FIG. 7.

[1] It is useful to adopt conventional symbols for the representation of
strata of different lithological characters, and so far as possible to
adhere to the same kind of symbol for any one kind of deposit. Those
which are generally in use, are rough pictorial representations of the
characters of the deposits, as shown in Fig. 7. The conglomerate is
indicated by circular marks representing cross-sections of the pebbles,
a breccia by triangular marks signifying that the fragments are angular
and not rounded; a sandstone is indicated by dots to represent the
grains of sand; a mud, clay or shale by continuous or broken horizontal
lines, which reproduce the appearance of the planes of lamination so
frequent in beds of this composition; a limestone is usually marked by
the use of regular horizontal lines illustrating the pronounced bedding,
with vertical lines at intervals to represent the regular jointing which
occurs in so many limestones: the nature of the bedding may be
further shown by drawing the lines comparatively far apart when the
limestone is a thick-bedded one, nearer together when it is thin-bedded.
Igneous rocks are represented by crosses or irregular V-shaped marks,
illustrating the absence of stratification and presence of joints.

Volcanic ashes are sometimes represented by dots, at other times
by signs somewhat similar to those which are used for true igneous
rocks. Sedimentary rocks which are composed of more than one kind of
material may be further shown by a combination of two symbols, thus

The horizontal section is one which is in constant use by the practical geologist : the results of the first traverse of a district may be jotted down in his note-book in the form of a horizontal section (with accompanying notes), and the written memoir on the geology of any district composed largely of stratified rocks will almost certainly require illustration by means of these sections. Perhaps nothing more clearly marks the careful observer than the nature of the sections which he makes, and geological literature is too frequently marred by the publication of slovenly sections. A badly drawn section not only offends the eye, it may and frequently does convey inaccurate information.

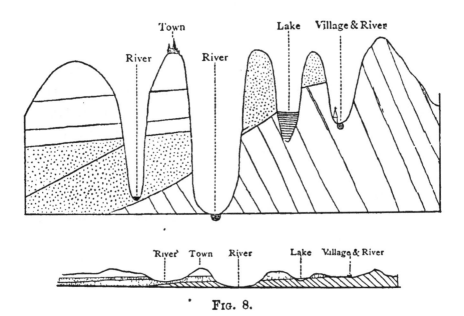

FIG. 8.

the existence of a sandy clay may be shown by means of a combination of horizontal lines and dots, and so with other combinations. The practical geologist should become accustomed to the use of these symbols in his note-book; if used, they will save much writing.

These symbols are used in some of the later illustrations to this book.

In the above figure (Fig. 8) taken from Sir Henry de
la Beche's " Sections and Views Illustrative of Geological
Phænomena,". Plate II., the lower drawing represents a
section drawn to true scale, while that above shows
one which is exaggerated. The student who saw this
would infer that the uppermost beds on the left side
of the upper section rested unconformably upon the
dotted beds beneath, and once abutted against them in
that portion of the figure where the beds have been
removed by denudation in the deep valley, whereas an
examination of the section drawn to true scale shows that
the unconformity does not exist (although there is one at
the base of the deposits marked by dots), and that there
is room for the higher deposits to pass above those marked
by dots at the place where the former have been removed
by denudation. Whenever possible, horizontal sections
should be drawn to true scale, the vertical heights being
on the same scale as the horizontal distances. Sections
which are so drawn represent the nature of the surface of
the country as well as the relationship of the strata, and
often illustrate in a marked degree the influence which
the character of the strata has exerted upon the nature of
the superficial features of a country. If it be impossible
to draw a section in which the elevations and horizontal
distances are represented upon a true scale, the former
ought to be drawn on a scale which is a multiple of
the latter; thus the vertical heights may be shown on
2, 3, or 4 or more times the scale chosen for the horizontal
distances; when this is done, it will often be necessary to
show the strata with an exaggerated dip, and accordingly
the exaggerated section loses some of its value, though if
vertical and horizontal scales bear some definite proportion
it will still be more valuable than a rough diagram which
is not drawn to any scale.

Section-drawing cannot be satisfactorily accomplished without some practice, and the student is strongly advised to acquire the art of drawing good sections; the writer can assert as the result of considerable experience in the conduct of examinations of all kinds, that slovenly sections are the rule in candidates' papers, and good sections very rarely appear. Study of the six-inch maps and horizontal sections (drawn on the same scale) of the Geological Survey of the United Kingdom will enable the student to familiarise himself with admirable sections, and it should be his aim to produce sections like these. He is recommended to take some of these six-inch maps which show contour-lines as well as the disposition of the strata, and to draw sections on the scale of six inches to the mile, vertical and horizontal, exhibiting the proper outline of the ground and the arrangement of the strata, and afterwards to compare them with the published sections. The sections should be drawn as far as possible at right angles to the general strike of the strata. Some datum-line is taken for the base of the section (say sea-level) and offsets drawn vertically from this where the section crosses a contour-line or recorded height. The height is marked on these offsets; thus if a recorded height of **2700** feet (just over half a mile) occurred on the line of section a height of somewhat over three inches is marked on the offset, and so with the other points where the section crosses contours or recorded heights. By joining these points on the offsets, giving the connecting lines curves similar to those which are likely to occur in nature, the general character of the surface of the ground is represented. The geology of the district is next shown. Wherever a dip is marked on the map, the direction and amount of dip is shown by a short line on the section, and where dips are not actually seen along the line of section, the dips which are nearest to that line on the map must be considered, and marked on the section. The lines of junction between the various deposits shown by different colours upon the map are inserted on the section as short lines, the inclination being judged by study of the nearest dips; faults and igneous rocks must be marked off, and any indication of the hade of the fault or the slope of the edges of the igneous rock which the map affords will be taken into account. The section will then appear somewhat as shown in the following figure:

Fault

Fɪɢ. 9.

and sufficient indication of the trend of the rocks will be obtained to shew that they form portions of curves which may then be filled in as shown in Fig. 10 and the section will be complete.

Fault

Fɪɢ. 10.

It will be noticed that the small dyke of igneous rock on the right of the main dyke is joined to it lower down, though no indication of this is given along the line of section; but the requisite information for this and evidence of the existence of the small dyke proceeding from the left-hand side of the main one may be obtained by the study of the rocks in a valley on one side or other of the line of section.

After the student has become conversant with the nature of geological maps and sections, and has read Sir A. Geikie's *Outlines of Field Geology*, he should on no account omit to learn something of the art of making geological maps, by going into the field and attempting to produce a map, for the art of geological surveying does not come naturally to any one, and some acquaintance with the methods of surveying is a necessity to everyone who wishes to make original geological observations, though all cannot expect to afford the time and acquire the skill necessary for the production of maps vying with the detailed maps of the Government Survey. Before actually attempting to draw lines on a map on his own account, he will do well to tramp over a portion of a district with the published geological map in his hands, selecting a country which is not characterised by great intricacy of geological structure, and he can then attempt to represent the geology of another portion of the same district without consulting the published map. Of all the districts of Britain with which he is acquainted the writer believes that the basin of the river Ribble, in the neighbourhood of the town of Settle in the West Riding of Yorkshire, is best adapted for studying field geology in the way suggested above, for the main geological features are marked by extreme simplicity, and the exposures are good, whilst the presence of an important fault-system and of a great unconformity relieve the area from monotony. Anyone who stands on the summit of Ingleborough or Penyghent will grasp the main features of a portion of the district without any difficulty, for it lies beneath his feet like a geological model, and when the student has mastered and mapped in the leading features, he can find bits of country with geology of varying degrees of complexity amongst the Lower Palæozoic rocks of the valleys which run down to Ingleton, Clapham, Austwick and Settle.

The biologist is supplied with laboratories at home and abroad, where he may study his science under the best conditions. Would that some munificent person would found, in a district like that referred to above, a geological station where Cambridge students would have the means of acquiring a knowledge of field-geology under conditions more favourable than those presented by the flats around the sluggish Cam!

CHAPTER IX.

EVIDENCES OF CONDITIONS UNDER WHICH STRATA WERE FORMED.

THE establishment of the order of succession of the strata, and the correlation of strata of different areas merely pave the way for the geologist. To write the history of the earth during various geological ages, he has to ascertain the physical and climatic conditions which prevailed during the successive geological periods, and to study the various problems connected with the life of each period. In the present chapter an attempt will be made to illustrate the methods which have been pursued in order to write to the fullest degree which is compatible with our present knowledge, the earth-history of various ages of the past. In making this attempt, the physical and climatic conditions may be first considered, and their consideration followed by that of the changes in the faunas, though it will frequently be necessary to refer to one set of conditions as illustrative of the other.

It will be assumed here that the great principle of geology, that the modern changes of the earth and its inhabitants are illustrative of past changes, is rigidly true. Reference will be made to this principle in a later chapter, but it is sufficient to state here that the study of the

sediments which have been deposited from the commencement of Lower Palæozoic times to the times in which we
now live bear the marks of having been formed under
physical conditions, which, in the main, are similar in kind
to those which prevail upon some part of the surface of the
lithosphere at the present day.

One of the most important inferences of the strati-
grapher relates to the existence of marine or terrestrial
conditions over an area at any particular time, and we
may, in the first place, consider the evidence which
supplies us with a clue to this subject.

It has been previously stated that the ocean is essentially the theatre of deposition, the land that of destruction,
and accordingly, the presence of deposit as a general rule
indicates the evidence of marine conditions during the
formation of those deposits, though this is not universally
the case.   Again, as denudation is practically confined to
the land areas, and the shallow-waters at their margins,
unconformity on a large scale gives evidence of the
existence of terrestrial conditions in the area in which
it is developed, during its production.   Accordingly a mass
of deposit separated from deposits above and below by
marked unconformities shows the alternation of terrestrial
conditions (during which the unconformity was produced)
and marine conditions (during which the deposits were
laid down).   The deposits formed after an unconformity
has been developed will naturally be of shallow-water
character, as will also be those of the period immediately
preceding the incoming of conditions which will cause the
occurrence of another unconformity, and between these
two shallow-water periods will occur a period when deeper-
water conditions probably prevailed.   We can therefore
not only divide the history of any particular area into a

series of chapters, of which every two successive ones will describe a continental period and a marine one, but each marine period may be divided into three phases—a shallow-water phase at the commencement, an intermediate deeper-water phase, and a shallow-water phase at the end. These phases are frequently complicated by the occurrence of a host of minor changes, but on eliminating these, the effects of the three great phases are shown by study of the nature of the strata, and their recognition does much to simplify the detailed study of the stratigraphical geology of various parts of the earth's surface.

In discriminating between terrestrial conditions and marine ones, the existence of unconformities is of great importance in marking terrestrial conditions and is often the only available evidence, for no accumulations or deposits formed on the land may be preserved to testify to the terrestrial conditions[1]. When terrestrial deposits and accumulations do occur, they are extremely important, and it is necessary to allude to the points wherein they differ from marine deposits.

Apart from organic contents, the mechanically formed deposits of rivers and lakes resemble in general characters the shallow-water deposits of the ocean, though they are usually less widely distributed. It is the accumulations which have actually been formed as æolian rocks, or those which have been laid down as chemical precipitates in salt-lakes which, by study of lithological characters, furnish the most convincing evidence of their terrestrial origin.

---

[1] The term terrestrial is used above in opposition to marine, to include the conditions prevalent above sea-level. The term continental would be better if it did not exclude insular conditions. Accordingly deposits formed in rivers, and fresh-water and salt-water lakes are spoken of as terrestrial.

Many æolian accumulations may be looked upon as soils, if the term soil be used in a special sense to refer to the accumulations which are produced as the result of the excess of disintegration over transportation in an area, whilst others are due to transport which has not been sufficiently effective to carry the material to the sea. When the weathered material accumulates above the weathered rock, it depends chiefly upon climate whether the disintegrated rock becomes mingled with much decayed organic matter forming humus. If this organic matter exists in quantity, the probability is that the accumulation is a terrestrial one, though this is by no means necessarily the case, for under exceptional circumstances a good deal of humus may be deposited in the sea, as beneath the mangrove-swamps which line the coasts of some regions, and to go further back, in the case of the Cromer Forest series of Pliocene times, or some coals, such as the Wigan Cannel Coal of the Carboniferous strata.

In addition to the work of water, which affects both land and sea-deposits, the land is especially characterised by the operations of wind and frost upon it, for these produce results which may frequently serve to differentiate a land-accumulation from a deposit laid down beneath sea-level. The effect of wind in rounding the grains of sand which are blown by it is well-known, and samples of the 'millet-seed' sands of desert regions are preserved in most museums. The greater rounding which characterises wind-borne as compared with water-borne sand grains is due, in great measure, to the greater friction between the grains when carried by the air than when swept along by the water. Under favourable circumstances water-worn grains may become rounded, especially when agitated by

gentle currents sweeping over a shoal[1]; but a large mass
of sand, in which most of the grains have undergone much
rounding so as to give rise to 'millet-seed' sand, will
nevertheless be probably formed by wind-action except
where a marine deposit is formed of material largely
derived from an earlier æolian one. The effect of frost is
to split rocks into fragments which are more or less
angular before they are subjected to water-action. The
broken fragments are prone to collect on slopes as screes,
and as any scree-material falling into the sea is likely
to become rounded except under conditions which rarely
prevail, the existence of much scree-material in a rock
suggests its terrestrial origin. Glaciers gave rise to terrestrial moraines, which may occasionally be identified as
land-accumulations by mere inspection of their physical
characters, but all geologists are aware of the difficulties
with which they are confronted when they attempt to
discriminate between terrestrial and marine glacial deposits.

The existence of much material amongst the stratified
rocks which has been precipitated from a state of solution
is an indication of the terrestrial origin of the rocks, which
were laid down on the floors of the inland seas, separated
more or less completely from the open ocean; for the
waters of the ocean are capable of retaining in solution
all of the material which is brought down to them, and
accordingly precipitates of carbonate of lime, rock-salt,
gypsum and other compounds formed from solution, are
only formed on a large scale in inland lakes, though they
may be formed to some extent when the water of a

---

[1] Cf. Hunt, A. R., "The Evidence of the Skerries Shoal on the
wearing of Fine Sands by Waves," *Trans. Devon. Assoc.*, 1887, vol. XIX.
p. 498.

lagoon is only slightly connected with that of the open ocean, and the evaporation is great, for instance in the lagoons of coral reefs. Certain physical features often mark the deposits of chemical origin, cubical or hopper-crystals of rock-salt may be dissolved, and the hollow afterwards filled with mud, so that the rock surfaces are sometimes marked with pseudomorphs of mud after rock-salt. Sun-cracks and rain-prints impressed on the rock are not actual indications of terrestrial origin of the rocks on which they are found, for the shallow-water muds of an estuary may be deposited in the sea and yet exposed to the action of the air at low tide, but they mark very shallow-water deposits which have been exposed to the atmosphere immediately after their formation if not during the time they were formed, and they frequently occur amongst the deposits of inland lakes.

It will be observed that the characters of the terrestrial accumulations serve to distinguish them to some extent from the marine ones, but they also enable one to detect to some degree the actual conditions under which the accumulation was produced, whether on the mountain-slope, or in the plain, the desert or the fen, the river-bank or the lake-floor.

The conditions of formation of the marine deposits may be distinguished within certain limits with ease, by examination of their physical characters, for the near-shore deposits will generally be coarser and contain more mechanically-transported material than the sediments which accumulate at a greater distance from the shore, though it is not safe to infer that deposits are formed away from the shore on account of the absence of mechanically-transported sediments. In districts where the mechanically-transported material is rapidly deposited, organic deposits of great purity may form close to the coast-line; for instance, when the rivers of a country end in fjords, the mechanical sediments are deposited in the fjords, and the sea around the coast is free from this sediment, and there the organisms can build up deposits of great purity; and a similar thing may happen when the rivers on one side of a country have short courses, and do not carry down much sediment, which occurs when the watershed is near the coast. On the one hand, clay may be formed in considerable purity near the coast, where the supply of mud is so great that the organisms existing there can do little in the way of contribution to the mass of the deposit, or it may be formed on the other hand in great depths of the ocean, where the supply of sediment is extremely small, but where all the organic tests become dissolved; as the characters of the deep sea clays are mainly negative, a geologist examining the rocks of the geological column would have much difficulty in distinguishing a deep-water clay from a shallow-water one by its lithological characters only. In cases of difficulty, information of importance is likely to be furnished by examination of the relative thickness of equivalent deposits in adjoining areas, for if we find a mass of clay a few feet thick in one region represented by hundreds of feet of clay and limestone in another, the former mass probably accumulated slowly and at some distance from the land; again, the uniformity of lithological characters of a deposit over a very wide area is a possible indication of its formation away from land, but this is not a safe guide, for reasons which will eventually appear, unless it can be shown that the deposit is everywhere of the same age.

A clue to climatic conditions is frequently furnished by the physical characters of accumulations, especially terrestrial ones.   The accumulations containing a large percentage of hydrocarbons have probably been formed under fairly temperate and moist climatic conditions, whilst the existence of millet-seed sandstones associated with chemical deposits points to desert conditions and inland lakes, requiring a dry climate and probably a warm one.   Glaciated surfaces and glacial deposits of course indicate a low temperature.   Some geologists profess that occasionally they can even determine the direction of the prevailing winds during past periods, by examination of the character of ripple-marks, rain-pits and other features, though it is doubtful whether much reliance can be placed upon these obscure indications.

Useful as is the physical evidence supplied by deposits, as an index to the conditions under which they were formed, it is usually only supplementary to the evidence derived from a study of the fossils.   Fossils when present in the rocks, usually supply considerable information concerning the prevalent conditions during the deposition of the rocks.   By them we can not only separate marine from terrestrial deposits, but also freshwater deposits from æolian accumulations; each kind of deposit will generally contain the remains of organisms which existed under the conditions prevalent in the area of formation of the rock, though it is of course a frequent thing for a terrestrial creature or plant to be washed into a freshwater area or into the sea.   In an æolian deposit, the invertebrate remains may be those of any air-breathing forms, as insects, galley-worms, spiders, scorpions and molluscs. The land-molluscs are all univalve.   Of vertebrates, we may find the bones and teeth of amphibians, reptiles, birds and mammals.   Occasionally freshwater or even marine forms may be found in an æolian deposit, but they will be exceptional. Marine shells are often blown amongst the sand-grains of the coastal dunes, and seagulls and other birds frequently carry marine organisms far inland.

The creatures frequenting fresh water differ from those of the land and of the sea. The most abundant vertebrate remains will be those of fishes, and of the invertebrates we find mollusca preponderate. The variety of molluscs is not so great as in the case of marine faunas. The bivalves always possess two muscular scars on each valve (except adult *Mulleria*); whilst many marine shells as the oyster have only one muscular scar on each valve. (See Fig. 11.)

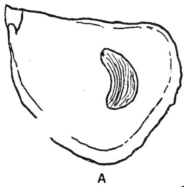
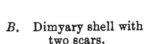

FIG. 11.

A. Monomyary shell with one scar.   *B.* Dimyary shell with two scars.

These scars mark the attachment of the adductor muscles, for drawing the valves together, and the shells with only one impression on each valve are called *monomyary*, those with two impressions *dimyary*. The discovery of monomyary shells indicates with tolerable certainty the marine character of the deposit in which they are found, though their absence cannot be taken as proof of freshwater origin. The beaks or umbones of the bivalves are often corroded in freshwater deposits, as may be seen by examining shells of the common freshwater mussel. "All univalve shells of land and freshwater species, with the exception of *Melanopsis* and *Achatina*, which has a slight indentation, have entire mouths; and this circumstance may often serve as a convenient rule for distinguishing freshwater from marine strata; since if any univalves occur of which the mouths are not entire, we may presume that the formation is marine[1]."

A

B

FIG. 12.

A. Holostomatous shell.   *B*.   Siphonostomatous shell.

In Fig. 12 *A* shows a freshwater shell (*Vivipara*) with entire mouth, whilst *B* exhibits the shell of a marine gastropod (*Pleurotoma*) with a notched mouth. The entire-mouthed shells are called *holostomatous* whilst those which are notched, the notch being often prolonged into a canal, are termed *siphonostomatous*.

Many groups of invertebrates are seldom or never found in fresh water. Of exclusively or nearly exclusively

[1] Lyell's *Students' Elements of Geology*, Second Edition (1874), Chap. III. A good account of the differences between freshwater and marine organisms, from which some of the facts here cited are extracted, will be there found.

marine creatures we may name the foraminifera, radiolaria, sponges with a hard framework, most hydrozoa
which secrete hard parts, corals, echinoderms, cirripedes,
king-crabs, locust-shrimps, most polyzoa, brachiopods,
pteropods, heteropods, and cephalopods. Of extinct
groups, the graptolites and trilobites seem to have been
entirely confined to the sea.

In the modern and comparatively modern deposits,
the forms frequently belong to existing genera, and we
get fairly conclusive evidence of the conditions of deposit
by determination of the genera. The terrestrial (including
freshwater) molluscs have mostly a long range in time.
We find pulmoniferous gastropods of living genera in the
Carboniferous period, one (*Dendropupa*) belongs to a subgenus of the modern land-shell *Pupa*, the other (*Zonites*)
to a subgenus of the snail group *Helix*. Many freshwater
molluscs as *Unio, Cyclas*, and *Physa* are found amongst
the secondary rocks, and give a clue to the origin of the
deposits which contain them. Many extinct genera are
closely allied to modern genera, and their mode of existence may be assumed with fair certainty. With all
these guides, we may sometimes be left in doubt as to the
conditions of deposit when organisms are few in number;
thus, it is yet a matter for discussion whether the Old
Red Sandstone and many of the deposits of the Coal
Measures of Britain were of freshwater or marine origin.

In considering the possibility of fossils having been
carried from land to water or *vice versa*, it will be remembered that generally speaking they are more readily
transferred from a higher to a lower level, so we are
more likely to find remains of land-animals and plants in
fresh water or the sea, and relics of freshwater animals
and plants in the sea, than of marine or freshwater animals

and plants in land, or marine organisms in fresh water. River-gravels and lacustrine deposits are especially prone to contain a considerable intermixture of land-forms with those proper to the station.

Fossils supply much information concerning the depth and distance from land at which the deposits were laid down.   When portions of the ocean-water have been separated to form inland lakes, the water becomes salter than that of the open ocean, if the evaporation is greater than the supply of fresh water, and the life of the inland sea undergoes change under the unfavourable conditions set up.   Many forms disappear altogether, and those which survive tend to become stunted, and the shells of many of the mollusca are abnormally thin; the fauna of an inland sea though it may have abundance of individuals is apt to be characterised by paucity of species.

Turning now to the faunas of the open oceans, it is found that in addition to latitude, the distribution of organisms is affected by depth, and by the nature of the sea-floor, and accordingly we find different organisms in different areas; and in examining the same area the organisms inhabiting different depths are not all the same, and at the same depth some kinds of animals have different *stations* from those of others, one creature being confined to a sandy floor, another to a muddy one, and so on [1].   The oceans have been divided into 18 *provinces*, each of which is more or less characterised by the possession of peculiar forms which are termed *endemic*, in contrast to the *sporadic* forms which are widely distributed.   In any area which is margined by a coast line,

---

[1] For an account of the distribution of one group of organisms see Woodward, S. P., *A Manual of the Mollusca*, from which many of the following observations are taken.

the molluscs are distributed in zones which were formerly classed as follows :—the *littoral* zone between tide marks, the *laminarian* zone from low water to fifteen fathoms, the *coralline* zone between fifteen and fifty fathoms, and the *deep-sea coral* zone from fifty fathoms to one hundred fathoms or more; this last depth was once supposed to mark the limit of the downward extension of marine life, but as the result of modern deep-sea soundings we know that organisms extend to a much greater depth, and the deep-sea fauna, owing to uniformity of conditions over wide areas, contains fewer endemic forms in proportion to the sporadic ones than the shallow-water[1]. The deep-sea deposits entomb the remains of these deep-sea organisms and also of numerous *pelagic* organisms which live upon the surface of the ocean, whose remains sink to the ocean-floor after death. Amongst the deposits of the deeper parts of the ocean, we find many which are almost exclusively composed of the tests of foraminifera, radiolaria and pteropods, the spicules of sponges, and the frustules of diatoms; and accordingly the existence of foraminiferal, pteropodan, radiolarian, and diatomaceous oozes, amongst the strata of the geological column, has been taken by some as indicating the prevalence of deep-sea conditions during the formation of those deposits: as the purity of a calcareous ooze depends upon the absence of mechanical sediment, or volcanic dust, and as the component organisms of these oozes are pelagic forms which live near the continents as well as in the open oceans, the presence of calcareous oozes implies the existence of a *clear* sea during their deposition but not necessarily of a deep one, for if the sea-area be far away from land masses, or if the sediment be strained off in

[1] For an account of the deep-sea fauna, see Hickson, S. J., *The Fauna of the Deep Sea*, 1894.

fjords, calcareous oozes may be formed in shallow water. The existence of pure radiolarian or diatomaceous deposits is better evidence of deep water, for if they were formed in shallow water we should expect an intermixture of calcareous tests, whereas these are dissolved whilst sinking into the extreme depths of the ocean. As the deep-sea creatures are under very different conditions from those of shallower waters, we might expect marked structural differences between the deep and shallow-water creatures: one such difference has been emphasized, namely the occurrence of animals which are blind or have enormously developed eyes in the great depths of the sea, where the only light is due to phosphorescent organisms. This is well seen in the case of many recent crustacea, and has been noted by Suess in the case of the trilobites of some beds which he accordingly infers to be of deep-water origin, and it is interesting to find that these creatures are found in deposits which give independent evidence of an open-water origin. The *Æglinæ* of the Ordovician strata are frequently furnished with enormous eyes, and they are often accompanied by blind trilobites, and in Bohemia the blind and large-eyed forms are sometimes different species of the same genus, for instance *Illænus*[1].

As one would naturally expect, the actual depth at which deposits were formed can generally be calculated with a greater degree of certainty amongst the newer rocks than amongst the older ones. In the case of the Pliocene Crags, the depth in fathoms may be confidently given. In the Cretaceous rocks attempts have been made to give numerical estimates of the depths at which different accumulations were formed, but some differences of opinion have arisen in the case of these rocks. In the

[1] Suess, E., *Das Antlitz der Erde*, 2ᵉˢ. Bd., p. 266.

Palæozoic rocks, only a rough idea of the general depth can usually be obtained, and no attempt to calculate the depth in fathoms is likely to be even approximately correct in the present state of our knowledge.

The comminution of fossils has sometimes been taken as an indication of shallower water origin of the deposits which contain them, but although the hard parts of organisms in a broken condition have frequently been shattered by the action of the waves, they may also be broken at great depths by predaceous creatures, and in many instances the fracture is the result of earth-movements occurring subsequently to the formation of the deposits.

Turning now to the difference in organisms which results from difference of station, it will be sufficient to give a quotation from Woodward's *Manual of the Mollusca* as an illustration :—" In Europe the characteristic genera of *rocky* shores are *Littorina, Patella,* and *Purpura ;* of sandy beaches, *Cardium, Tellina, Solen ;* gravelly shores, *Mytilus ;* and on muddy shores, *Lutraria* and *Pullastra.* On rocky coasts are also found many species of *Haliotis, Siphonaria, Fissurella,* and *Trochus ;* they occur at various levels, some only at the high-water line, others in a middle zone, or at the verge of low-water. *Cypræa* and *Conus* shelter under coral-blocks, and *Cerithium, Terebra, Natica* and *Pyramidella* bury in sand at low-water, but may be found by tracing the marks of their long burrows (Macgillivray)[1]."

The geologist will naturally select sporadic forms rather than endemic ones in comparing the strata of different areas, but how far differences in faunas are the result of existence at different times, and how far they

---

[1] Woodward, S. P., *A Manual of the Mollusca,* p. 151.

are due to difference of conditions affecting contemporaneous organisms can only be discovered as the result of accurate observation. The main points to be regarded when comparing the successive faunas of different regions have been noticed in this and the preceding chapters, and it has been shown that as the evidence is cumulative, it requires the collection of a large number of facts obtained by observation of the strata before accurate inferences can be drawn.

The indications of climatic conditions furnished by organisms require some consideration. In the comparatively recent deposits it is not difficult to get some notion of the prevalent climatic conditions when the fossils belong to forms closely related to modern genera. The existence of the arctic birch and arctic willow, and of shells belonging to species now living north of the British Isles, in deposits of comparatively recent date in Britain would afford convincing evidence of the occurrence of colder climatic conditions than those which are now prevalent in the area, even if the evidence were not confirmed as it is, by physical proof of glaciation in deposits of the same age. Nevertheless, even in these recent beds, we have a useful warning, by finding species of elephant and rhinoceros associated with northern forms like the lemming, glutton, and musk-ox. We know that the species of elephant and rhinoceros (the mammoth and woolly rhinoceros) were provided with thick coverings which would enable them to resist the severity of an arctic climate, but had not these coverings been found, we might have been puzzled by the association of forms whose nearest allies are subtropical with others of arctic character. As we go back in time and deal with earlier deposits, the ascertainment of the climatic conditions becomes more difficult, as the fossils mostly belong to extinct species, genera or even families.

In these circumstances, it is very dangerous to draw conclusions as to climatic conditions from examination of a few forms, but when we find that plants and animals, terrestrial and marine forms, vertebrates and invertebrates alike point.to the same conclusion, as in the London Clay, where all the fossils belong to forms allied to those now living under sub-tropical conditions, the state of the climate may be inferred with considerable certainty[1]. The character of the fossils must be taken into account rather than their size. There was a tendency amongst geologists to believe that large organisms probably indicate warm conditions. Recent researches in arctic seas have dispelled this belief. Marine algæ of enormous size are found in the cold seas, and the size of creatures, abundance of individuals and variety of forms in the arctic faunas of some regions is very noteworthy. In the Kara Sea, for instance, a variety of creatures were dredged up during the voyage of the Vega, and Baron Nordenskjöld makes the following pertinent remarks about them: "For the science of our time, which so often places the origin of a northern form in the south, and *vice versa*, as the foundation of very wide theoretical conclusions, a knowledge of the types which can live by turns in nearly fresh water of a temperature of + 10°, and in water cooled down to − 2°·7 and of nearly the same salinity as that of the Mediterranean, must have a certain interest. The most remarkable were, according to Dr Stuxberg, the following: a species of Mysis, *Diastylis Rathkei* Kr., *Idothea entomon* Lin., *Idothea Sabinei* Kr., two species of Lysianassida,

[1] For a discussion as to the value of plants as indices of climate see Seward, A. C., Sedgwick Essay for 1892.

Pontoporeia setosa Stbrg., *Halimedon brevicalcar* Goë
an Annelid, a Molgula, *Yoldia intermedia* M. Sars, *Yoldi*
(?) *arctica* Gray, and a Solecurtus[1]." The temperature
were taken by a centigrade thermometer. Again we rea
of the results of dredging off Cape Chelyuskin. "Th
yield of the trawling was extraordinarily abundant; larg
asterids, crinoids, sponges, holothuria, a gigantic sea
spider (Pycnogonid), masses of worms, crustacea, etc. *1
was the most abundant yield that the trawl-net at any on
time brought up during the whole of our voyage round th
coast of Asia*, and this from the sea off the norther
extremity of that continent[2]."

Amongst the marine invertebrates reef-building coral
and mollusca perhaps furnish the best evidence of climati
conditions. The coral-reefs of the Jurassic rocks wit
large gastropods and lamellibranchs clustered around ther
have been appealed to in proof of the existence of sub
tropical conditions during their formation; further bac
in time we find evidence of climate furnished by th
fossils of the Silurian rocks of the Isle of Gothland in th
Baltic Sea. Of these, Lindström writes "*The fauna had
tropical character*. In consideration of the great number
of Pleurotomariae, Trochi, Turbinidae and the larg
Pteropods the assumption of a tropical character of th
fauna may seem justifiable[3]."

Structure may give some indication of climate eve
though the organism is not allied to living species. Th
bark of trees in arctic regions is often thicker than i
more temperate regions, and the leaves of arctic plant

---

[1] Nordenskjöld, A. E., *The Voyage of the Vega*, Vol. I. Chap. IV.

[2] *Ibid.* Chap. VII.

[3] Lindström, G., *On the Silurian Gastropoda and Pteropoda of Go
land*, Stockholm, 1884, p. 33.

often have special characters to enable them to resist the long periods during which they are deprived of water, though the fact that desert-plants frequently shew similar modifications deprives this test of any particular value except as a means of corroborating conclusions reached from other evidence[1]. The shells of arctic mollusca may become stunted, but this is not by any means universal, and the same result may be brought about by other abnormal conditions, as for instance the increase of salt in a water area by evaporation.

On the whole, an examination of the evidence available for ascertaining the character of climate by reference to included organisms, shews that inferences may be drawn within certain limits, but that the task is a difficult one not unaccompanied by danger, and every kind of available evidence derived from a study of physical phenomena and the included organisms should be utilised before any conclusion is drawn.

The likelihood of accurate inference is increased by comparing the faunas of various areas; should they seem to indicate a progressive lowering of climate when passing from lower to higher latitudes, it is probable that the indication is correct. The student is referred to a paper by the late Professor Neumayr for an account of the existence of climatic zones during the Mesozoic Period[2].

[1] For an account of the modifications of the leaves of arctic plants, see Warming, Eug., *Om Grønlands Vegetation*, Meddelelser om Grønland, 12th part, p. 105.

[2] Neumayr, M., "Ueber klimatische Zonen während der Jura- und Kreidezeit," *Denkschrift. der Math.-Naturwissensch. Classe der k. Akad. der Wissenschaften*, Bd. XLVII. Vienna, 1883.

CHAPTER X.

EVIDENCES OF CONDITIONS UNDER WHICH STRATA WERE FORMED, CONTINUED.

IN the preceding chapter, attention was drawn to the indications as to conditions of deposition furnished by the sediments of any one locality, and only passing reference was made to variation in the nature of the sediments and their organic contents, when the deposits are traced laterally from place to place; some attention must now be paid to this matter.

It is sometimes inferred that, whereas similarity of organisms is a dangerous guide in correlating the strata of two areas, accurate correlations may be made, if the deposits can be traced continuously through the intervening interval; no doubt the task is simplified when this can be done, but the continuity of deposit of one particular composition is no more proof of contemporaneity than the occurrence of the same fossils continuously through the interval, imbedded in strata of different character, indeed probably not so much so. The existence of wide-spread masses of conglomerate, which are not found as linear strips, but which extend in all directions, is in itself an indication of this; the Oldhaven pebble bed for instance, in the Tertiary rocks of the London basin, is very widely distributed. We cannot suppose that coastal conditions prevailed far away from the shore-line, and accordingly when a conglomerate occurs in a wide-spread sheet, and not in a linear strip, this is indicative that the deposit has not been formed continuously but that strip has been added to strip along an advancing or receding shore line, and if this happens with conglomerates, it must occur also in the case of other deposits.

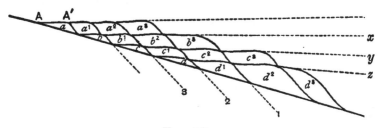

Fig. 13.

In fig. 13[1] let A represent a shore line of a continent which is undergoing gradual elevation. A deposit of pebbles a will be formed against the coast, one of sand b further away, then one of mud c and lastly limestone d, may be formed in the open sea away from land. Naturally there may be intermingling of two kinds of deposit at the junctions, but for the sake of simplicity this may be disregarded. During the accumulation of the deposits a, b, c, d, certain sporadic forms may be distributed throughout all the deposits, and some of them may become extinct before the deposition of these beds is completed, if the process is carried out on a large scale; we may speak of the characteristic fossils of this period as fauna I. As the result of elevation or of mere silting up of the sea-margin, or of both combined, the next mass of pebble-deposit will be laid down further away from

[1] The writer gratefully acknowledges his indebtedness to Prof. Lapworth for some of his views concerning deposition of strata.

the original shore, for the shore line will now be at A' and not at A, and it will partly overlap the mass of sand b; the sand b^1 will also be deposited somewhat further out and partly overlap the mud c, and similarly the mud c^1 will partly overlie the limestone d. During the formation of a^1, b^1, c^1, d^1, other sporadic forms belonging to a fauna II may replace those of the first fauna. In the same way a^2, b^2, c^2, d^2 will be deposited, and in the meantime a new fauna III may arise and replace II. So the process will go on until we finally have a group of deposits lying one over the other, consisting of a basal accumulation of limestone, succeeded by mud, sandstone and pebble-beds in succession. Each of these will be continuous, though the inner part of the pebble-deposit was formed long before the outer part of the limestone, which is nevertheless beneath a mass of pebble-deposit continuous with that formed first, and the various deposits will be separated by fairly horizontal planes x, y, z, which might be regarded as bedding planes, but which are not so, strictly speaking. The true bedding planes will occur at a slight angle to these planes of separation, for the structure resembles false bedding on a gigantic scale, but of course, the lines separating two masses of similar deposit will be practically horizontal and parallel to the planes of demarcation of two distinct kinds of material. The lines separating two faunas would, under the conditions postulated, run approximately parallel to the planes of separation of adjoining deposits of the same lithological character but would pass from conglomerate, through sandstone, mud and limestone, as indicated by the lines 1, 2, 3, ... and the deposits between adjoining lines would be contemporaneous[1]. In nature, complications will arise, owing to

[1] The lines 1, 2, 3 ... are incorrectly drawn in the figure. Line 1

the gradual appearance and disappearance of forms, and the existence of endemic species in contemporaneous deposits formed in different stations and having different lithological characters.

If elevation ceased and were succeeded by depression, the exact opposite would occur, and the pebble beds would be overlain by sandstones, these by muds, and lastly limestones would appear. It follows that during a marine phase occurring between two unconformities we should have a **V**-shaped accumulation of deposits with the apex pointing to the part of the shore line which was last submerged before the commencement of elevation, as shewn in fig. 14, though the beds of the apex will in most cases be denuded during the re-emergence.

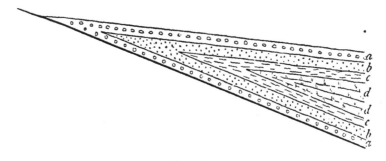

Fig. 14.

Indications of the non-coincidence of the planes separating faunas and those which separate deposits of one lithological character from those of another have already been detected, for instance the 'greensand' condition of the Cretaceous period occurs in some places during the existence of one fauna, and in others during that of another, though the planes have not been traced

should be drawn so as to separate a, b, c, d from a^1, b^1, c^1, d^1, line 2 to separate a^1, b^1, c^1, d^1 from a^2, b^2, c^2, d^2, and so with the others.

continuously. Mr Lamplugh has furnished another example amongst the Cretaceous rocks of Yorkshire and Lincolnshire, but as has already been observed, a great deal remains to be done in this direction, and geologists are much in want of two sets of stratigraphical maps, in one of which the lines are drawn with reference to the differences of lithological character, whilst in the other they separate different faunas.

The student will notice the normal recurrence of deposits in definite order; conglomerate succeeded by sandstone, mud and limestone, in a sinking area, and limestone succeeded by mud, sandstone and conglomerate in a rising area. Naturally many instances of departure from this rule are seen, owing to local conditions, but on a large scale, it is very frequently noted, and recognition of this will enable the student to remember the variations in the lithological characters of the deposits more easily, than if he simply acquired them from a text-book without taking heed as to their significance.

Upon the variations in the lithological characters of deposits and of their faunas, when the beds are traced laterally depends very largely the successful ascertainment of the existence of former coast-lines, the restoration of which constitutes an important part of Palæo-physiography, concerning which some observations may here be made[1]. If a set of deposits having different lithological characters can be proved to be contemporaneous, the coarser detrital accumulations will point to the approach to a coast-line, and the actual position of the coast during the period of accumulation of the deposits may be very accurately fixed. The pebble-beds at the base of the

[1] On this subject, the student may consult Prof. E. Hull's *Contributions to the Physical History of the British Isles.*

Cambrian rocks of Llanberis indicate the existence of a coast-line in that position during the accumulation of those pebble-beds. Similar pebble-beds occur at St David's, at the base of the Cambrian, but it is impossible in the case of these rapidly accumulated sediments to say that two deposited so far away from one another were actually contemporaneous, and therefore although we might draw a line through Llanberis and St David's to indicate the old coast-line of the period, it does not follow that the actual beach existed simultaneously at the positions indicated. The palæo-physiographer, however, attempts to restore the physical conditions of greater thicknesses of deposit; for instance, the distribution of land and sea during Lower Carboniferous times over the area now occupied by the British Isles is often taken to illustrate the methods of restoration of ancient features, and all admit that the lithological and palæontological characters of the rocks indicate a shallowing of the Carboniferous sea when passing northwards towards Scotland. For conveying an idea of the restorations to the student, it is almost imperative to portray the distribution of land and sea upon a map, and this can only be done by drawing definite lines. It must be distinctly understood that these lines are necessarily only an approximation to the actual position of the ancient shore-lines, which must have shifted again and again during the long period occupied by the accumulation of the Lower Carboniferous strata, so that a true idea of the positions of the Lower Carboniferous shore-lines could only be obtained by placing on a series of maps the successive shore-lines of different parts of the Lower Carboniferous period, and taking a composite photograph of these, which would appear as a wide belt of shaded portion of the map with no definite boundaries. The utmost that the maker of palæo-physiographical maps can expect to indicate, when dealing with considerable thicknesses of strata, is an approximation to the mean position of the shore-lines of the period when these strata were deposited. This is extremely valuable in enabling the student to understand the significance of the variations in the characters of the strata and their organic contents, if he distinctly recognises the generalised nature of the map. Examination of any two palæo-physiographical maps of the same period by different authors will shew wide divergences in the details, but a general resemblance of the main features. The reader will do well to consult Prof. Hull's restoration of the physical features of Old Red Sandstone and Lower Carboniferous Times on Plate VI. of his *Contributions to the Physical History of the British Isles*, and compare it with the map drawn by Prof. Green (*Coal: its History and Uses*, by Profs. Green, Miall, Thorpe, Rücker, and Marshall, Fig. 3, p. 38), which will be found to bear out this statement.

Valuable as the published maps of palæo-physiography are as an aid to the student in understanding the significance of the variations of characters amongst the sediments, he will do well to supplement them by maps which he fills in for himself. He is recommended to procure a number of outline maps of England, or of the British Isles, and when studying in detail the characters of the British sedimentary rocks formed during the various periods, to place a blank map by his side when beginning the study of each period or important portion of a period. On this map he should jot down the geographical distribution of the different kinds of sediments, using the conventional signs indicated at p. 90 : thus, in the case of

the Lower Carboniferous rocks he would place the conventional sign for limestone in Derbyshire, a combination
of those for limestone and shale in Yorkshire, and would
add to these the sandstone sign in Northumberland. He
should also note the general character of the fossils, using
abbreviations for such terms as fresh-water fossils, shallow-
sea fossils, deep-water fossils. After reading the account
of the group of rocks in a comprehensive text-book, and
inserting his notes on the map, he should proceed to insert the probable position of the coast-lines. He should
also take notes of any indications of contemporaneous
volcanic action, though these might well be inserted on a
separate map. If this course be pursued, the student will
not only have the significance of the variations amongst
the strata impressed upon his mind, but he will have a
means of obtaining at a glance the distribution of
sediments and faunas of different kinds in the British area
during the principal geological periods. On another set
of maps he may indicate the axes of the orogenic movements which have occurred at different times, and when
his various maps are completed, he will have the materials
for the construction of a general account of the various
geological processes which have been concerned with the
building of the British area.

When an area like Britain has been studied, the
student may proceed to construction of maps of wider
regions, and he will find that in doing this, new sets of
facts must be taken into consideration, as for instance the
occurrence of different faunas on opposite sides of once-
existing continental masses, and the problems connected
with the present distribution of the faunas and floras.
For an instance of the importance of the former distribution of life the reader may consult the twelfth section of

the first part of Professor Suess' *Das Antlitz der Erde,* whilst a good account of the value of recent geographical distribution of organisms in supplying a clue to former distribution of land and sea will be found in Mr A. R. Wallace's *Island Life,* Chapter xxii.

Should the method suggested above be adopted, the student is likely to acquire a much more coherent idea of the significance of the facts of stratigraphical geology than can be obtained by a mere perusal of the accounts of the strata given in those portions of the various text-books which are devoted to a consideration of the stratigraphical branch of the science.

CHAPTER XI.

THE CLASSIFICATION OF THE STRATIFIED ROCKS.

In the succeeding chapters, a general account of the characters of the Geological Deposits of different periods will be given, for the purposes of illustrating the principles to the consideration of which the earlier chapters have been devoted. It is not proposed to enter into a description of numberless details, which would only confuse the student who wished to grasp the main principles, for many facts have been recorded which it is necessary to notice in a comprehensive text-book treating of stratigraphical geology, though their full significance is not yet grasped. The writer, while noting the main characters of the various subdivisions of the different stratigraphical systems, will assume that this work is used in conjunction with some recognised text-book. The stratigraphical portion of Sir A. Geikie's *Class Book of Geology* gives an admirable general account of the British Strata, while the larger text-book by the same author has a condensed though very full account of the rocks of the stratigraphical column in all parts of the world, and this is supplemented by numerous references to the original works wherein further descriptions may be found. The English edition of Prof. E. Kayser's *Text-Book of*

Comparative Geology, edited by P. Lake, is also we
adapted to the wants of the student, and an excellen
account of the strata is given in Mr A. J. Jukes-Browne'
Handbook of Historical Geology, which may be read wit
the same author's *Building of the British Isles*.

The reader who refers to different text-books will b
struck with the variations of nomenclature even amongs
the larger stratigraphical divisions, for two authors seldon
subdivide the geological column into the same number c
rock-systems. The following classification will be her
adopted :—

Groups.	Systems.
Cainozoic or Tertiary	Recent Pleistocene Pliocene Miocene Oligocene Eocene
Mesozoic or Secondary	Cretaceous Jurassic Triassic
Palæozoic	Permian Permo-Carboniferous Carboniferous Devonian Silurian Ordovician Cambrian.
Precambrian.	

A few remarks may be given as to the reason fo
adopting this classification.

It is not for a moment suggested that the System
have the same value, if the time taken for their accumula
tion be alone considered. The beds classified as Recent

for example, were probably accumulated during a lapse of time far shorter than that occupied for the deposit of some of the series or even stages of a system like the Silurian, but the recent rocks acquire a special significance from the fact that we are living in the period, and the Cainozoic rocks as a whole are capable of greater subdivision than the earlier groups, on account of the greater ease with which they can be studied, owing to the small amount of disturbance which they have usually undergone when compared with that which has affected older rocks, and the closer resemblance of their faunas and floras to those of existing times.

With reference to the groups, the writer has already commented upon the use of the terms Palæozoic, Mesozoic and Cainozoic; below the lowest Palæozoic rocks (those of the Cambrian system) lie a group of rocks which have been variously spoken of as Azoic, Eozoic, and Archæan. There is an objection to the use of any one of these words in this sense; the objection in the case of the first two is that the term is theoretical and probably incorrect, whilst the word Archæan, otherwise suitable, has also been used in a more restricted sense. In these circumstances the term Precambrian will be used when referring to any rocks which were formed below Palæozoic times, though no doubt when this obscure group of rocks is more thoroughly understood a satisfactory classification will be applied to it.

Taking the other groups into account, the lower systems of the Palæozoic group will be found to vary greatly according to the views of different writers; some make only one system, the Silurian, others two, the Cambrian and Silurian. The three systems are here adopted, not only because the one, Silurian, is too unwieldy on account of its size and requires subdivision (and the Cambrian and Silurian however defined, will be found to be of very unequal importance, whereas the three systems adopted are of fairly equal value), but especially because when the term *Ordovician* is used the significance of the other terms Cambrian and Silurian is at once understood.

An attempt has been made to shew that the Devonian system is non-existent, but the result of modern research is to shew that the rocks placed in this system are worthy of the distinction, both from their importance and from the distinctness of the fauna from those of the underlying and overlying systems.

The Fermo-Carboniferous system is adopted, because an important group of deposits has recently been brought to light which were not represented either in the Permian or Carboniferous system as originally defined.

Some authors have advocated the union of the Permian and Triassic systems into one system placed at the base of the Mesozoic group. This is unnecessary, and would depart from the classification originally proposed, which is to be deprecated, unless there is any strong reason for it.

The Mesozoic systems are classified according to the method generally adopted. Were a fresh classification to be proposed, a portion of the Cretaceous system might be included with the Jurassic rocks, but it is better to adhere to the old classification.

The divisions of the Cainozoic rocks are hardly system in the sense in which the term is used in the case of the older rocks, but the reason for using these smaller subdivisions has already been mentioned. The addition of the Oligocene to the original divisions suggested by Lyell has been found useful, and the term will be used in this work.

The reasons for the adoption of the particular minor subdivisions (series and stages) in the following chapters will frequently appear when the rocks of the various systems are described, and need not be further alluded to in this place.

Although most geologists describe the stratified rocks in ascending sequence beginning with the oldest, and proceeding towards the newest, others, and notably Lyell, adopted the opposite method and commenced with an account of the newest beds. The argument generally used for the latter method is that it is easier to work from the study of the known to that of the less known, and as the faunas of the newest rocks are most like the existing faunas, the student would more readily follow a description of the rocks in the order which is opposite to that in which they were deposited.

In practice, the study of the sediments in their proper order, that is, in the order of deposit, will not be found to task the student to any great extent, especially if, as is very desirable, he has studied the main facts and principles of Palæontology before commencing the study of the rock-systems in detail. There is one reason for beginning with the study of the older sediments which outweighs any reasons which can be advanced against it, namely that the events of any period produce their effect not only upon the strata of that period, but also on those of succeeding periods.

The task of the stratigraphical geologist is really to learn the evolution of the earth, in its changes from the simple to the more complex conditions, and it is quite obvious that it is unnatural to attempt any study of evolution by working backward. For this reason the study of the sediments will be here made in the order which is usually adopted, by passing from the older to the newer, and from the simple to the more complex.

The British strata will be mainly considered, though references will frequently be made to their foreign equivalents, and a fuller account of the latter will be added when the British strata are abnormal, as are those of Triassic times, and also when a period is not represented amongst the strata of the British Isles, as for instance, the Fermo-Carboniferous and Miocene periods.

The student is recommended to refer constantly to good geological maps of the British Isles, of Europe, and of the world. Of maps of the British Isles, mention may be made of Sir A. Ramsay's geological map of England, Sir A. Geikie's map of Scotland, and his map of the British Isles, J. G. Goodchild's map of England and Wales, a map of Europe by W. Topley and one of the world reduced from that by J. Marcou, accompanying the first and second volumes of the late Sir J. Prestwich's *Geology*. For special purposes more detailed maps will be studied, including the one-inch maps of H. M. Geological Survey, and the index map on a smaller scale. Lastly, for an account of British Geology, reference must be made to H. B. Woodward's *Geology of England and Wales*, where the British formations are described in order, and to W. J. Harrison's *Geology of the Counties of England and Wales*, where the stratigraphical geology of the country is given under the head of the different counties, which are taken in alphabetical order.

In concluding this chapter, it is hardly necessary to say that every opportunity of studying the characters of the deposits and their fossils in the field should be eagerly seized, and that much information may be acquired even on a railway journey, especially as to the influence which the deposits exert upon the scenery of a region[1].

[1] In the first edition of H. B. Woodward's *Geology of England and Wales*, an account of the geology of the main lines of English railways is given, which is omitted in the later edition. It is well worth consulting by those who take a long journey, and it will be found useful to take a geological map with one on the journey so as to discover when one is passing from one formation to another.

CHAPTER XII.

THE PRECAMBRIAN ROCKS.

STUDY of a geological map of the world will shew that extensive regions, such as parts of Scandinavia, many tracts of Central Europe, a large area in Canada, and a considerable portion of Brazil and the adjoining countries are occupied by crystalline schists, which underlie the oldest known sedimentary strata in those places. These crystalline schists form the floor upon which the sediments constituting the bulk of the geological column rest, and it is necessary that we should know something of the character of this floor. Other rocks which can be definitely proved to be of Precambrian age are often found associated with the crystalline schists, and these associated rocks have often undergone more or less alteration subsequently to their formation. The difference between the coarser types of crystalline schists and these associated rocks is sometimes so marked that geologists have necessarily paid attention to it, and separated the two groups of rocks; the term Archæan has been used by some geologists to include the crystalline schists, and Eparchæan for the associated rocks of known Precambrian age, but though this separation may sometimes be effected, there are cases when it is impossible to draw any sharp line of demarcation between 'Archæan' and 'Eparchæan' types.

In the present state of our knowledge, a chronological classification of the Precambrian rocks when applied to wide and distant regions is destined to break down, and it will be convenient if we consider at some length the features of the Precambrian rocks of a particular region, and apply the knowledge thus gained to a study of Precambrian rocks of other areas, and to a consideration of our knowledge of the Precambrian rocks as a whole. In doing so, the term 'crystalline schists' will be used somewhat vaguely with reference to a complex of schistose rocks of which the mode of origin cannot be fully determined. We may take our own country as a region where a good development of the Precambrian rocks occurs.

A few explanatory remarks concerning the mode of detection of Precambrian rocks may not be amiss. If any true organisms have been hitherto discovered amongst the rocks formed before Cambrian times they are valueless as a means of correlating rocks, and accordingly lithological characters only are available in attempting to correlate the rocks of one area with those of another. Those who have read the preceding chapters will have gathered that comparisons founded on similarity of lithological character are not so valuable as those made after careful scrutiny of the fossils of strata, but they are by no means valueless, and when the rocks of two areas which are not far distant from one another present close lithological resemblances, their general contemporaneity may be inferred with some degree of certainty.

It is only when we get the lowest Cambrian strata overlying earlier rocks that we have absolute proof of the Precambrian age of the latter, and it is necessary, therefore, that we should have some definite lower limit to the rocks of the Cambrian system. It is now generally agreed that that limit shall be drawn at the base of a group of rocks containing what is known as the *Olenellus*-fauna, which will be considered at greater length in the next chapter, and it will be well, if the term Cambrian be not in future applied to any rocks beneath the ones containing the relics of this fauna, for otherwise there is danger of the indefinite downward extension of the Cambrian system. We need not be surprised to find great thicknesses of rock below the rocks containing the *Olenellus*-fauna, and passing upwards with complete conformity into those rocks; nevertheless, if it can be shewn that the *Olenellus*-fauna had not appeared during the deposition of the underlying group, the rocks of that group should be termed Precambrian. A case of this nature has not yet been detected in our area, and all the rocks which have been proved to be Precambrian in Britain are separated from the overlying Cambrian rocks by a physical break, though that break is not necessarily very large, and in some districts is probably of little importance. Hitherto the *Olenellus*-fauna has been detected in Ross, Warwickshire, Shropshire, Worcestershire and probably in Pembrokeshire, and the rocks underlying the *Olenellus*-beds in those counties can be proved to be Precambrian (i.e. if the *Olenellus*-age of the Pembrokeshire rocks be ultimately established, and the researches of Dr Hicks tend to prove that it will almost certainly be done). It will be convenient if we take the instances where the age of the rocks can be proved with certainty or with a considerable degree of probability first, and then consider the examples of rocks which are found below Cambrian strata, though these have not hitherto yielded the *Olenellus*-fauna, concluding with a notice of rocks which have been claimed to be of Precambrian age on account of their lithological characters, though they are not now seen to be immediately succeeded by strata appertaining to the Cambrian system.

Commencing with the region where we have the greatest development of the known Precambrian rocks, namely Ross, Sutherland and the Hebrides, we may explain the general relationship of the rocks by means of a generalised section (fig. 15).

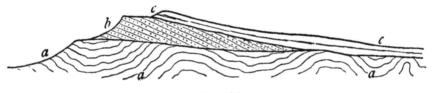

FIG. 15.

The lowest rocks *a* are crystalline schists, they are succeeded by a set of arenaceous rocks *b* known as the Torridonian beds, which rest unconformably upon the upturned edges of the crystalline schists, whilst the Cambrian rocks, *c*, rest with another unconformity sometimes upon the partly denuded Torridonian beds, or where the latter have been completely removed, as on the right side of the figure, directly upon the crystalline schists, thus presenting an example of unconformable overlap. The occurrence of the *Olenellus*-fauna in the basement beds of the Cambrian system near Loch Maree, proves the Precambrian age of the Torridonian strata, whilst the unconformable junction between the latter and the crystalline schists indicates that we are here dealing with two distinct sets of Precambrian rocks, one of Eparchæan and the other of Archæan type.

The crystalline schists consist of rocks of very varied lithological characters, some with gneissose, and others with schistose structure, and they vary in degree of acidity from ultrabasic rocks to those of acid composition. Most of them exhibit parallel structures, which in many cases can be shewn to have been impressed on the rocks subsequently to their consolidation, though this need not have occurred and probably did not occur with some of them, especially the granitoid gneisses. The researches of the members of H. M. Geological Survey have shewn that many of these rocks were originally intrusive igneous rocks, though it is not yet known into what rocks those which were first consolidated were injected, and the origin of the bulk of the schists still remains to be elucidated. Subsequently to their consolidation and before the deposition of the earliest Torridonian rocks they were subjected to more than one set of earth-movements, which folded them and impressed a series of parallel structures upon many of them; and accordingly we find that the pebbles of the crystalline schists which are found amongst the basal conglomerates of the Torridonian rocks consist of fragments which had undergone the alteration caused by these earth-movements before they were denuded from their parent-rocks[1].

The Torridonian system is composed of rocks which are largely of arenaceous character, the most prominent beds being formed of red sandstones, and the bulk of the fragments in them have clearly been derived by denudation from the crystalline schists, many of the beds being

---

[1] For an account of these rocks, their characters, and the effects of earth movement upon them, the reader should consult a "Report on the Recent Work of the Geological Survey in the North-West Highlands of Scotland": *Quart. Journ. Geol. Soc.*, vol. XLIV. p. 378.

composed of arkose, where the quartz is mixed with a large proportion of felspar and often of ferro-magnesian minerals. The deposits are clearly sedimentary, and are as little altered as many strata of much more recent origin, only possessing structures produced by metamorphic action under exceptional circumstances. The detailed researches of the geological surveyors prove that the rocks of this system have a much greater thickness and are of more varied lithological characters than was previously supposed. The total thickness of the strata is over 10,000 feet, and the sandstones are associated with deposits of a muddy character, and with occasional bands of limestone; in these circumstances the discovery of fossils would excite no surprise, and in 1891 Sir A. Geikie announced the detection of "traces of annelids and some more obscure remains of other organisms in these strata," which have not yet been described[1]. These Torridonian strata furnish us with the most satisfactory group of Precambrian sediments yet detected in Britain[2].

In the south-east Highlands is a great mass of crystalline schists of a less gneissose character than that of the north-west, to which Sir A. Geikie has applied the name Dalradian. Many of these schists will be found by examination of the geological map of Scotland to be separable into divisions, which by means of their lithological characters can be traced long distances across the country, and they present all the characters of sedimentary rocks, though they are associated with intrusive igneous rocks,

---

[1] An account of the subdivisions and lithological characters of the rocks of the Torridonian System will be found in the *Annual Report of the Geological Survey of the United Kingdom* for 1893.

[2] It has been recently maintained that some of the Torridonian rocks are of Æolian origin.

and have undergone great metamorphic changes since their formation. Cambrian rocks have not yet been discovered immediately above them, though they are clearly older than Ordovician times, but the existence of rocks associated with them along their north-west borders, which in lithological characters closely resemble some of the rocks of the crystalline schists of the north-west Highlands, indicates the probability of their general Precambrian age. In some instances, the extreme types of metamorphism which they exhibit are the result of the kind of action usually termed pyrometamorphic as has been shewn by Mr G. Barrow[1].

In England and Wales the rocks which have been shewn or inferred to be Precambrian, when not intrusive, are largely of volcanic origin. The most satisfactory example of the occurrence of the *Olenellus*-fauna is that of the Cambrian Comley sandstone of Shropshire, which rests unconformably upon a set of rocks termed by Dr Callaway the Uriconian rocks; the latter are essentially volcanic, and strongly resemble Precambrian rocks of other British areas. There is also strong reason to suppose that the sediments to which the name Longmyndian has been applied, which have been described by the Rev. J. F. Blake, are of Precambrian age, for, as Professor Lapworth has pointed out, the three great sub-divisions of the Cambrian system are present in the area under consideration, and the rocks of each are entirely different from those of the adjoining Longmynd area. In Shropshire therefore we meet with one set of volcanic rocks and another set consisting of sedimentary rocks, of which

[1] Barrow, G. "On an Intrusion of Muscovite-biotite gneiss in the S.E. Highlands of Scotland, and its accompanying metamorphism. *Quart. Journ. Geol. Soc.*, vol. XLIX. p. 330.

the former is certainly, the latter almost certainly of Precambrian age, and as the Longmyndian rocks are in a comparatively unaltered condition, consisting of normal sediments, we may well expect the discovery of fossils in them also[1]. The *Olenellus*-fauna has been found near Nuneaton in Warwickshire in beds which unconformably succeed volcanic rocks, the Caldecote series of Prof. Lapworth, and the latter are therefore of Precambrian age[2]. A few fossils belonging to the *Olenellus*-fauna have occurred in the oldest Cambrian rocks of the Malvern district, and these rocks rest unconformably upon those of an old ridge which is therefore composed of Precambrian rocks. The rocks of this ridge are largely of intrusive igneous origin, though parallel structures have been impressed upon them as the result of subsequent deformation, but some of the rocks are almost certainly of contemporaneous volcanic origin[3]. In the Wrekin ridge, igneous and pyroclastic rocks are found succeeded unconformably by Cambrian rocks which resemble those

[1] The reader may consult a paper by Prof. Lapworth "On *Olenellus Callavei* and its geological relationships," *Geol. Mag.* dec. III. vol. VIII. p. 529, for information concerning the relationship of the *Olenellus* beds of Shropshire to the more ancient rocks; the Uriconian rocks are described by Dr Callaway in a series of papers, especially in the *Quarterly Journal of the Geological Society*, vol. XXXV. p. 643, vol. XXXVIII. p. 119, vol. XLII. p. 481 and vol. XLVII. p. 109, whilst the lithological characters of the Longmyndian rocks are described by the Rev. J. F. Blake (*Quart. Journ. Geol. Soc.*, vol. XLVI. p. 386).

[2] See Lapworth, C., "On the sequence and systematic position of the Cambrian rocks of Nuneaton," *Geol. Mag.* dec. III. vol. III. p. 319; and Waller, T. H., "Preliminary Note on the Volcanic and Associated Rocks of the neighbourhood of Nuneaton," *ibid.* p. 322.

[3] For details concerning the rocks of the Malvern Hills see papers by Callaway in the *Quarterly Journal of the Geological Society*, vol. XXXVI. p. 536, XLIII. p. 525, XLV. p. 475, and XLIX. p. 398, and a paper by Prof. A. H. Green, *ibid.* vol. LVI. p. 1.

of the Malvern and Nuneaton districts, and probably belong to the period of existence of the *Olenellus*-fauna, and these igneous and pyroclastic rocks are presumably of Precambrian age, and the contemporaneous rocks constitute Dr Callaway's typical Uriconian group. Volcanic ashes and breccias are accompanied by devitrified pitchstones and intruded granitic rocks, which may or may not be all of the same general age[1]. The rocks which have been claimed as Precambrian in Pembrokeshire and in Caernarvonshire have the same general characters as those of the Wrekin ridge. Pyroclastic rocks underlie the oldest Cambrian rocks, with discordance between the two, and associated with these pyroclastic rocks are quartz felsites which according to some are of contemporaneous nature whilst others maintain their intrusive origin. In each county granites are found which are now generally recognised to be intrusive, though there seems to be no doubt as to their being of the same general age as the rocks with which they are associated, and therefore presumably Precambrian. The Pembrokeshire rocks are marked by the occurrence of a certain amount of metamorphism, probably of more than one kind, which has converted pyroclastic volcanic rocks into sericitic-schists and quartz-felsites into hälleflintas[2]. The term Pebidian

[1] Callaway, C., *Quart. Journ. Geol. Soc.*, vol. xxxv. p. 643.

[2] The Pembrokeshire area is of interest as the probable existence of Precambrian rocks in Britain was first indicated on good evidence in this county. The general structure of the district is fairly simple, consisting of Cambrian rocks beneath which Precambrian rocks are exposed in at least two ridges of which the northerly and more important one runs through St Davids. The rocks of the St Davids ridge consist of a binary granite (granitoidite), felsites, and volcanic ashes and breccias of intermediate composition. Much diversity of opinion has existed, and to some extent still exists as to questions of detail, and a very extensive literature has been devoted to these rocks. Amongst the numerous

given by Dr Hicks to the contemporaneous volcanic fragmental rocks should be retained, and if these rocks be eventually shewn to be contemporaneous with similar volcanic rocks of other districts, may be applied generally, as it has priority over other terms as Uriconian and Caldecote series. The term Dimetian was applied to rocks known to be intrusive, and must be dropped as a chronological term, whilst the existence of an Arvonian system separate from the Pebidian system is not fully proved.

In Caernarvonshire two ridges are found, the one running from Bangor to Caernarvon, and the other through Llanberis lake. The rocks of these are generally similar to those of St Davids, and as the lowest Cambrian rocks of the area closely resemble those of St Davids, the Precambrian age of the rocks of these ridges is rendered highly probable, though until the discovery of the *Olenellus*-fauna in the area, it cannot be regarded as proved [1].

The actual position of the similar rocks of Anglesey has not been so clearly fixed, as the rocks associated with them are of Ordovician age, but their resemblance to the rocks of the adjoining regions renders their Precambrian age highly probable. It is interesting to find in associa-

papers which treat of them, the student may consult the following :— Hicks, H., *Quart. Journ. Geol. Soc.*, vol. xxxiii. p. 229, xxxiv. p. 147, xxxv. p. 285, xl. p. 507, xlii. p. 351, Geikie, A., *ibid.* vol. xxxiv. p. 261, Blake, J. F., *ibid.* vol. xl. p. 294, and Morgan, C. Ll., *ibid.* vol. xlvi. p. 241. Much of the matter contained in these papers is controversial, and need not be fully read by those who merely wish to obtain a general account of the rocks of the district.

[1] These rocks are described by T. M^cK. Hughes, *Quart. Journ. Geol. Soc.*, vol. xxxiv. p. 137, and xxxv. p. 682 ; by Prof. T. G. Bonney, *ibid.* vol. xxxiv. p. 144 ; and by Dr Hicks, *ibid.* vol. xxxv. p. 295.

tion with the rocks which resemble those of Caernarvonshire, others which Sir A. Geikie recognises as quite
similar to some existing amongst the crystalline schists
of the north-west Highlands of Scotland, and when these
ancient rocks of Anglesey have been mapped in detail,
they will probably be found to present greater variety
than is afforded by any Precambrian rocks of Great
Britain occurring S. of the Scotch border[1].

Of rocks whose age is more uncertain, but which are
probably of Precambrian age, those of Charnwood Forest
in Leicestershire may first be noticed. They are largely
of pyroclastic origin, and from their likeness to similar
rocks of proved Precambrian age, they are very probably
of this age, as suggested by Messrs Hill and Bonney[2].
A group of crystalline schists is found in the south of
Cornwall, especially near the Lizard, and similar rocks
are found in the Channel Isles. As their relationship to
newer rocks is not clear, little can be said about them,
which has not already been noticed in mentioning the
crystalline schists of other regions[3].

The Precambrian rocks of the European continent
consist largely of crystalline schists which in their general
aspects recall those of the north-west Highlands of
Scotland. Important masses are found in Bavaria,

[1] Papers upon the old rocks of Anglesey will be found in many
volumes of the *Quarterly Journal of the Geological Society*; see especially
Hicks, vol. xxxv. p. 295, Callaway, vol. xxxvi. p. 536, xxxvii. p. 210, and
Blake, xliv. p. 463.

[2] Hill and Bonney, *Quart. Journ. Geol. Soc.*, vol. xxxiii. p. 754,
xxxiv. p. 199 and xlvii. p. 78 ; see also Watts, W. W., *Rep. Brit. Assoc.*
for 1896, p. 795.

[3] For an account of the Volcanic History of Britain in Precambrian
times, see Sir A. Geikie, Presidential Address to the Geological Society,
Quart. Journ. Geol. Soc., vol. xlvii. p. 63.

Bohemia, France, Spain, Scandinavia and Russia. The Scandinavian and Russian rocks of Archæan type are in places succeeded by the *Olenellus*-bearing beds of the Cambrian rocks, and rocks of Eparchæan character are not extensively developed, though certain Norwegian rocks may be the equivalents of the Torridonian rocks of Scotland, and other rocks of this type are found in places in Sweden. In Bohemia and in Brittany Precambrian strata of Eparchæan type have been discovered, and this type probably occurs elsewhere in Europe.

The North American rocks require some notice, for it was in Canada that the existence of Precambrian rocks was first recognised, and the term Laurentian, originally applied to an Archæan type of Precambrian rocks in Canada, was subsequently adopted in speaking of many Precambrian rocks elsewhere, though it is now wisely restricted to the type of rock in the original area to which the name was first given. These Laurentian rocks acquired a special interest on account of the occurrence in their limestones of a supposed reef-building foraminifer, *Eozoon canadense*, but detailed study of its structure and mode of occurrence has convinced most geologists that the structure is inorganic.

The Laurentian rocks of the typical Laurentide region are largely crystalline schists associated with massive crystalline rocks. The attempt to separate them chronologically into a Lower and Upper division was premature, as shewn by the fact that many of them, upon detailed study, prove to be intrusive igneous rocks. In the neighbourhood of Lake Huron, a set of sedimentary rocks overlying the Archæan rocks is of Eparchæan type, consisting to a great extent of volcanic rocks, clay-slates and schists with intrusive igneous rocks; it has been termed the Huronian System, and this term has also been extensively applied to other Eparchæan types found elsewhere, but should be restricted to the rocks of the Huron district. A number of other rocks of Eparchæan type have been discovered in various parts of North America, and have been grouped together under the title of Algonkian, a name proposed for them by Dr C. D. Walcott, and an attempt has been made to arrange them in chronological order, though in the absence of fossils, the rocks of different districts can only be so arranged by reference to lithological characters; nevertheless a detailed study of the Eparchæan and some of the more finely crystalline schistose rocks points to the existence of a number of divisions of sedimentary rocks of Precambrian age, some of which may attain to the dignity of forming separate systems[1]. By far the most instructive development of American Precambrian rocks has been found in the Rainy Lake region of Canada, and it is the subject of a special memoir by Dr A. C. Lawson[2]. The Archæan rocks of the region are divided into a lower Laurentian and an upper division, which is further subdivided into the Coutchiching series below and the Keewatin series above, though the rocks of the Keewatin series are largely of Eparchæan character. The Laurentian rocks of this region resemble those of the Laurentide area, and consist of highly crystalline schistose and gneissose rocks associated with compact rocks. The Coutchiching series consists of mica schists and grey

---

[1] A large number of classifications have been proposed for the Archæan rocks of America ; the most plausible one is given in Sir A. Geikie's *Text Book of Geology*, Third Edition, p. 716.

[2] Lawson, A. C., *Report on the Geology of the Rainy Lake Region.* Montreal, 1888.

laminated gneisses, which appear to have been of sedimentary origin, altered by subsequent metamorphic action,
while the Keewatin series, which reposes sometimes upon
the rocks of the Coutchiching series (when the junction
is an unconformable one), sometimes upon the Laurentian
rocks, is formed of pyroclastic rocks and lava flows with
intercalated sedimentary rocks; some of the Keewatin
rocks are highly metamorphosed but others have undergone little or no metamorphic change. The most important point in connexion with these rocks of the Rainy
Lake Region has reference to the relationship between the
Laurentian rocks and those of the Coutchiching and
Keewatin series. Lawson demonstrates the igneous nature
of the Laurentian rocks, and brings forward evidence of
various kinds that they were formed " by the fusion of the
basement or floor upon which the formations of the upper
division of the Archæan were originally deposited. With
the fusion of this floor it seems probable that portions of
the superincumbent strata, which once formed integral
parts of either the Coutchiching series or the Keewatin,
have also been absorbed into the general magma, and
reappeared on crystallization as Laurentian gneiss. This
fusion, however, only extended up to a certain uneven
surface, which surface constitutes the demarcation between
the present upper and lower Archæan. Above this surface,
or upper limit of fusion, the formation of the Coutchiching
and Keewatin series retained their stratiform or bedded disposition, and rested as a crust of hard and brittle rocks upon
the magma, subject to its metamorphosing influences[1]."

We may now pass briefly in review the evidence
which has been so far obtained as to the mode of formation of the various Precambrian rocks. The existence

[1] Lawson, *op. cit.* p. 139.

of a very varied fauna amongst the earliest Cambrian strata has been commented upon by many geologists, and according to accepted explanations of the origin of that fauna, an enormous period of time elapsed before the deposition of the earliest Cambrian strata. During portions of that long period, the undoubtedly clastic rocks of Eparchæan type were deposited, and probably many others which are now so altered by metamorphism, like some of the Coutchiching rocks of Canada, that their original clastic origin can only be inferred and not directly proved. Volcanic activity was very rife during the deposition of some of these Eparchæan rocks, though perhaps not more so than during the formation of some of the Lower Palæozoic Rocks. All attempts to prove the occurrence of organisms in Precambrian strata have hitherto failed, for no undoubted fossil has been described which is unhesitatingly accepted as of Precambrian age, notwithstanding the many asserted occurrences of such fossils. That fossils will eventually be discovered is more than probable, and their non-detection at the present time is in no way very surprising, when we remember the long time that elapsed after the existence of stratified rocks below the Upper Palæozoic rocks had been recognised, before definite faunas were discovered in them. The determination of the Precambrian age of stratified rocks is recent, and now that this determination has been made, the search for fossils will be more eager, and is likely to be rewarded by their discovery. Furthermore, experience shows that when fossils are discovered in rocks of unknown age, there is a tendency to refer those rocks to some known period, and consequently we may actually possess Precambrian fossils, out of beds which have been erroneously referred to the Cambrian or a later period.

Another important question is that of the metamerphism of a large number of Precambrian rocks, and here again recent research tends to show that the metamorphism is not of a kind different from that which occurred after the end of Precambrian times; the discovery of crystalline schists in Norway, Kirkcudbrightshire and Westmorland amongst Lower Palæozoic rocks, which resemble those of Archæan masses in all respects except in the extent of area which they cover, shows that similar processes to those which occurred in Precambrian times went on during later periods, though perhaps not on so large a scale. The great extent of these metamorphic rocks of Precambrian age can hardly be due in any great degree to the longer time during which they have been subjected to metamorphic influence, for there is evidence that much of the change took place in Precambrian times, far more than has occurred since, and it is a significant fact that these old rocks are more extensively penetrated by intrusive igneous masses than those of later periods; here again we find that much of the intrusion actually occurred in Precambrian times. The greater extent of intrusion and metamorphism amongst these Precambrian rocks than amongst later sediments indicates some differences of conditions in the case of Precambrian and later times. If besides intrusion, actual fusion of floors of Precambrian rocks occurred, we may well suppose that the earlier records of the rocks are for ever lost to us, the earliest sediments having been fused, but that the history of life upon our earth is to be revealed to us first in so late a stage as that of Cambrian times is highly improbable, and we may look forward with confidence to laying bare the records of the rocks composing the geological column some way below the Cambrian portion of the column.

Upon this foundation of igneous rock, sediment and volcanic material, formed in Precambrian times, whose history we have only begun to study, was laid down the great mass of sediment which the geologist has more completely studied, where abundant traces of life are preserved, and concerning whose history we can gain a greater insight than is permitted us in the case of the old Foundation Stones.

CHAPTER XIII.

CYCLES OF CHANGE IN THE BRITISH AREA.

BEFORE studying in further detail the strata of the geological column, it will be convenient to deal with the great physical changes which have occurred in the British area from Precambrian times to the present day, as this will clear the way for a right appreciation of the main variations in the characters and distribution of the strata.

At the end of Precambrian times there was a general upheaval of the British area, and this we may speak of as the First Continental Period. It was followed by depression and extensive sedimentation, proceeding more or less continuously though with local interruptions through Lower Palæozoic times, so that so far as Britain is concerned we may speak of Lower Palæozoic times as constituting the First Marine Period. Extensive upheaval gave rise to continental tracts and mountain chains, and deposits of abnormal character (as compared with ordinary marine deposits) at the end of Lower Palæozoic times;—the Devonian period was one of elevation and denudation, and we may therefore refer to it as the Second Continental Period. This was followed by depression and sedimentation in Carboniferous times, and these Carboniferous times constitute the Second Marine

Period. Elevation gave rise to continental tracts and mountain chains at the end of Carboniferous times, and here again we find proofs of extensive denudation and the formation of abnormal deposits :—the Fermo-Triassic period is the Third Continental Period. Depression set in during early Jurassic times and continued throughout the Mesozoic and the early part of Tertiary times, which form the Third Marine Period. Disturbances culminating in Miocene times once more produced terrestrial conditions. In this, the Fourth Continental Period, we are still living.

From what has been previously written it will be seen that each of the marine periods should be marked by an early and late shallow-water phase, separated by an intervening marine phase, and the importance of the phases will depend upon the length of time during which they existed, and will differ markedly in different cases, whilst the distinctness of the middle phase from the upper and lower, will depend upon the magnitude of the maximum submergence.

During the first marine period submergence was comparatively rapid, and the shallow-water phase only lasted through very early Cambrian times in most regions, whilst the deep-water phase, complicated by many minor upheavals, extended through the main part of Cambrian, Ordovician and Silurian times, and was replaced by the later shallow-water phase at the end of Silurian times.

The second marine period again was ushered in by rapid submergence, so that the shallow-water phase was brief, and the main mass of the Lower Carboniferous strata was deposited in deep water ; but, unlike the first marine period, the second was characterised by the occurrence of a long interval of time marking the later shallow-water phase, during which the whole of the Upper Carboniferous strata were deposited. The Carboniferous Marine Period is the simplest of the three with which we have to deal, as the local oscillations occurring on a fairly large scale for such movements were less frequent than was the case during the first and third marine periods.

The third marine period had a long shallow-water phase at the commencement, with many minor oscillations, causing great variation in the character of the deposits and frequent minor unconformities. This shallow-water phase existed throughout Jurassic and Lower Cretaceous times. The deep-water phase existed during the deposition of the Upper Cretaceous deposits, and was succeeded by the second shallow-water phase, when the early Tertiary strata were accumulated.

The difference between the elevations which accompanied the Continental Periods and those which have been alluded to as minor elevations is no doubt one of degree, but in considering the British strata only no confusion is likely to arise on this account, as the difference was here very great.

The events which occurred during the continental periods are of extreme importance to the geologist. Every great upheaval was accompanied by crumpling and stiffening of portions of the earth's crust, and a definite trend was given to the strata as the result of these movements. It is to the earth-movements of the four great continental periods that the present structure of the British Isles is largely due, and in any attempt to restore the physical history of our islands considerable attention must be paid to the changes which were produced in the stratified rocks during these periods of earth-movement.

CHAPTER XIV.

THE CAMBRIAN SYSTEM.

Classification. The rocks of the Cambrian system when found reposing on Precambrian rocks in Britain are always separated from the latter by an unconformity. The typical development of the rocks of the system, as the name implies, is in the hilly region of Caernarvonshire and Merionethshire in North Wales, and they are also well represented in South Wales, the border counties between England and Wales, and the North-West Highlands of Scotland. Two distinct classifications of the Cambrian rocks of Britain are in use, the original one founded on variations of lithological character, whilst the second depends upon faunistic differences, but the original lithological classification has been to some extent modified to make it locally correspond with the classification based upon palæontological grounds. The following table will shew the differences :—

Lithological Classification.	Palæontological Classification.
Tremadoc Slate Series[1]	Beds with Intermediate Fauna
Lingula Flags Series	Beds with *Olenus* Fauna
Menevian beds (formerly included in Lingula Flags)	Beds with *Paradoxides* Fauna
Solva beds ⎫ Formerly grouped together as Harlech or Llanberis beds	
Caerfai beds ⎭	Beds with *Olenellus* Fauna

[1] In accordance with the custom usually observed in Britain, the

The original lithological classification was essentially the result of Prof. Sedgwick's work in North Wales, while the classification according to faunas is the outcome of the researches of Dr Hicks in South Wales.

Description of the Strata. The Cambrian rocks of North Wales occur in two complex anticlines, separated by an intermediate syncline of Ordovician strata occupying the Snowdonian hills. The southerly or Harlech anticline forms a part of Merionethshire to the east of Harlech, whilst the northern one is developed around Bangor and Llanberis. The South Welsh Cambrian rocks are chiefly found on either side of the Pembrokeshire axis of Precambrian rocks which runs through St David's. As the corresponding rocks of the two regions were deposited in bathymetrical zones of much the same depth, it will be convenient to give a general account of the rocks of the two regions at the same time, leaving the student to acquire information of the detailed variations in the larger text-books and in special memoirs[1].

Tremadoc slates are placed in the Cambrian system; most continental geologists place them in the succeeding Ordovician system. The matter is not an important one, as the fauna is an intermediate one between that of the Lingula Flags and that of the Arenig series of the Ordovician system, and the beds are true beds of passage. As the lithological classification is essentially British, it will be as well to retain the Tremadoc Slates in the Cambrian system.

[1] A general account of the Cambrian, Ordovician and Silurian rocks will be found in the Sedgwick Essay for 1883, *A Classification of the Cambrian and Silurian Rocks*, though the use of a cumbrous nomenclature therein will tend to confuse the reader. For a detailed account of the Cambrian rocks of North Wales the reader is referred to the Geological Survey Memoir, *The Geology of North Wales*, by Sir A. Ramsay (2nd edition), he may also consult Belt, T., "On the Lingula Flags or Festiniog Group of the Dolgelly district," *Geol. Mag.*, dec. i. vol. iv. pp. 493, 536, vol. v. p. 5. The geology of the Cambrian rocks is described in a series of Memoirs in the *Quarterly Journal of the Geological Society*

The strata of the Caerfai and Solva groups show the prevalence of the shallow-water phase almost uninterruptedly through the whole of the time occupied by their accumulation in the Welsh areas. They consist chiefly of basal conglomerates, succeeded by alternations of grits and shales, though the latter are often converted into slates, owing to the subsequent production of cleavage. The basal conglomerates of the Caerfai beds are frequently marked by the existence of enormous pebbles, composed of fragments of the rocks of the underlying Precambrian groups, and the possibility of the occurrence of glacial action during their accumulation as advocated by Dr Hicks must be taken into account. Above these beds are various coloured grits, with alternations of muddy sediments often coloured red[1]. The Solva group consists of massive grits, of various colours, also with alternations of mud, which have prevalent purple and green hues. The great thickness of the strata of the Caerfai and Solva Series, which sometimes exceeds 10,000 feet, must also be noted.

The Menevian beds consist essentially of very fine, well laminated black and grey muds, which are of a

by Dr H. Hicks; the following should be consulted: Harkness, R. and Hicks, H., "On the Ancient Rocks of the St David's Promontory, South Wales, and their Fossil Contents," vol. xxvii. p. 384; Hicks, H., "On some Undescribed Fossils from the Menevian Group," vol. xxviii. p. 173; and "On the Tremadoc Rocks in the neighbourhood of St David's, South Wales, and their Fossil Contents," vol. xxix. p. 39. See also Hicks, "The Classification of the Eozoic and Lower Palæozoic Rocks of the British Isles," *Popular Science Review*, New Series, vol. v., and Hicks, "Life-zones in the Lower Palæozoic Rocks," *Geol. Mag.* dec. iv. vol. i. pp. 368, 399 and 441.

[1] In giving this description the red (Glyn) slates of North Wales are treated as belonging to the Caerfai series, though this correlation depends on lithological characters only at present.

texture favourable for the production of a somewhat regular jointing, causing the rock to break into small rectangular blocks. They are thin, not exceeding 600 feet in thickness, and indicate the in coming of the general deep-water phase of the Lower Palæozoic epoch. The Lingula Flags mark a local return to shallower water conditions, especially in the central portion. The total thickness is over 3,000 feet, of which the lower stage (locally the Maentwrog series) is over 500 feet, and consists of blackish muds, the middle (Festiniog stage[1]) is about 2,000 feet thick, and is composed chiefly of shallower water gritty flags, whilst the upper (Dolgelly) stage is of about the same thickness as the lower stage and has similar lithological characters.

The Tremadoc Slates are about 1,000 feet thick. They are divided into a lower and upper stage, of about equal thickness, and are essentially composed of iron-stained slates, with a considerable admixture of calcareous matter in some parts of South Wales, when they furnish the nearest approach to a limestone which has been found amongst the Welsh Cambrian strata. They were probably formed in a fairly deep sea.

Much pyroclastic rock and some lava flows are intercalated amongst the Welsh Cambrian sediments. Tuffs are formed in the lower beds of St David's, and lavas and ashes have been found amongst the Lingula Flags and Tremadoc Slates of North Wales, while the Lingula Flags of South Wales have furnished several bands of ash to the north of Haverfordwest. Much of the material of the grits and muds may be derived from

---

[1] The term Festiniog has been used for the whole Lingula Flag series as well as for the middle stage. It will be well to use it with reference to the stage only.

volcanic rocks, though how far this is so cannot be stated in the absence of information obtained by detailed petrological examination of the rocks.

The various isolated outcrops of Cambrian strata amongst the counties of the Welsh borders and adjoining Midland counties indicate a great thinning of the Cambrian rocks in this direction.

The probable equivalents of the Caerfai rocks occur at Nuneaton, Comley, and on the flanks of the Wrekin and Malvern hills. The thin basal conglomerates are succeeded by quartzites, and sometimes red calcareous sandstones (Comley sandstone). These rocks are succeeded by thin arenaceous and calcareous beds which represent either the Solva or Menevian beds of Wales. The Lingula Flags are represented by the Malvern Shales of the Malvern area and the Stockingford Shales of Nuneaton, whilst the Tremadoc Slates have as their equivalents the Shineton Shales. The exact thicknesses of these deposits do not seem to have been recorded, but Prof. Lapworth observes that in central Shropshire "the Comley and Shineton groups which...have a collective thickness of perhaps less than 3,000 feet, we have apparently a condensed epitome of the entire Cambrian system as at present generally defined."

The Cambrian rocks of the North-west Highlands consist of a thin conglomerate succeeded by grits and flags with shaley beds, and above these a mass of limestone, which may represent some of the Ordovician deposits as well as those of Cambrian age. Pending a complete description of the faunas of these rocks, it is sufficient to state that the only fauna which has hitherto been described in detail indicates the existence of Lowest Cambrian rocks. Further remarks will be made on this head when describing the character of the Cambrian faunas. The Cambrian rocks of the North-west Highlands are also very thin as compared with those of Wales, so that the Highland and Welsh borderland regions appear to have existed as a deeper sea area than that which is indicated by the Cambrian rocks of Wales, an inference which is to some extent borne out by study of the Cambrian rocks of extra-British areas, to which we may now turn.

The principal European developments of Cambrian rock are found in Scandinavia, Russia, Bohemia and Spain, and of these the Scandinavian one is by far the most fully developed, as there is a complete sequence in the rocks of that peninsula. They occur both in Norway and Sweden, but the Swedish exposures are the most interesting in most respects, especially those of Westrogothia and Scania. The rocks are of no great thickness, and consist essentially of black carbonaceous shales, with inconstant bands of impure black limestone composed almost entirely of the remains of trilobites or more rarely of brachiopods. These Alum Shales, as they are termed, rest unconformably upon Precambrian rocks, and have arenaceous and conglomeratic deposits at the base. In Russia the rocks are still further attenuated, and have not yielded the relics of so many faunas as have been found in the Scandinavian Cambrian rocks.

The Bohemian development is incomplete, owing apparently to an unconformity at the base of the overlying Ordovician rocks, while the Spanish deposits which seem fairly thick and composed largely of mechanical sediments have not been worked out in very great detail.

The American development of Cambrian rocks resembles the European one in many striking particulars,

and as in the case of Europe, there are lateral variations
in the lithological characters of the rocks, though in the
opposite direction, the shallow-water deposits occurring on
the east coast, and the deep-water deposits further west.

The general distribution of the different types of
Cambrian strata in Europe and North America has been
accounted for on the supposition that in Cambrian times
a tract of land lay over much of the present site of the
North Atlantic Ocean, and that the detritus of that land
formed the shallow-water accumulations of Wales and the
east of Canada, whilst further away from it were deposited
the open-sea accumulations of Scandinavia and Russia on
one side and of the more westerly regions of North
America on the other, as indicated in Fig. 16.

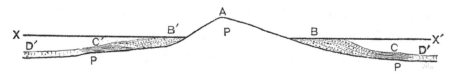

FIG. 16.

P. Precambrian Rocks.   BB'. Shore deposits.
A. Land.   CC'. Deep-water deposits.
X, X'. Sea level.   DD. Abyssal deposits.

The Cambrian Faunas.   The Cambrian Period has
been termed the age of trilobites, for they are the
dominant forms of the time, but they are associated with
many other forms of invertebrata; indeed all the great
groups of this division are represented in the earliest
Cambrian fauna. Dr C. D. Walcott records representatives of Spongiae, Hydrozoa, Echinodermata, Annelida,
Brachiopoda, Lamellibranchiata, Gastropoda, Pteropoda,
Crustacea and Trilobita as occurring in the *Olenellus* beds
of North America and other groups are represented in the
rocks of this age in the Old World. The Cambrian

trilobites as a whole are of more generalised types than those of the later systems which furnish their remains, as indicated especially by the looseness of the body, and the large number of body rings in many of the genera, while the tail or pygidium was small and formed of only a few coalesced segments, as pointed out by Barrande. In the later trilobites the test is more compact, there are on the whole fewer body rings, as more of these have become fused into a tail which is therefore larger than that of the average tail of the Cambrian trilobite.

Taking the faunas in order, the oldest or *Olenellus* fauna has furnished a great variety of forms in the Northwest Highlands of Scotland, Shropshire, Scandinavia, Esthonia, Sardinia, Canada, and Newfoundland, whilst representative species of the fauna have been recorded also from Worcestershire, Warwickshire, Pembrokeshire, India, China, and Australia.

The dominant form is the trilobite of the genus or group *Olenellus*, which contains a great variety of species referable to three or four divisions which have been ranked as separate genera by some writers. Associated with *Olenellus* are trilobites belonging to other genera, which are found in higher deposits, though there represented by different species.

Brachiopods are fairly abundant, especially those provided with a horny shell; of these, the genus *Kutorgina* is widely distributed.

The zoological relationships of several of the fossils of this horizon are as yet doubtful. The Archæocyathinæ show affinities with certain corals; a number of tests, included in the genus *Hyolithes* and its allies are doubtfully referred to the Pteropods, and the position of the genus *Volborthella* is uncertain. Special attention is directed to these doubtful relationships, as it is possible that a number of 'generalised forms' of organisms occur in these strata[1].

It should be noticed here that faunas have been discovered which are possibly of earlier date than the *Olenellus* fauna, as they do not correspond with it, or with those of newer strata. One, the *Neobolus* fauna of the Salt Range of India, occurs in beds below those with *Olenellus*, though it is not yet clear that *Olenellus* will not be eventually discovered associated with it, whilst the other, the *Protolenus* fauna of Canada, is of unknown age[2].

The *Olenellus* beds are succeeded by beds containing the *Paradoxides* fauna, which have been found in North and South Wales, Shropshire, Scandinavia, Bohemia, Spain, and North and South America. *Olenellus* and its allies became extinct (or else so scarce that no relics of them have been discovered in the *Paradoxides* beds) before the commencement of the deposition of the strata containing the *Paradoxides* fauna, and few genera pass from the beds with the one fauna to that containing the other. The *Paradoxides* fauna existed for a considerable period, and the beds have been divided into a series of zones characterised by different species of *Paradoxides*, thus

[1] For an account of the *Olenellus* fauna see Walcott, C. D., "The Fauna of the Lower Cambrian or Olenellus Zone," *Tenth Annual Report of the Director of the United States Geological Survey*, Washington, 1890. It is possible that some of the fossils mentioned in that report belong to strata above that containing *Olenellus*.

[2] For an account of the *Neobolus* beds see Noetling, F., "On the Cambrian Formation of the Eastern Salt Range," *Records Geol. Survey, India*, vol. xxvii. p. 71, and for the Protolenus fauna consult a paper by Matthew, G. F., "The *Protolenus* Fauna," *Trans. New York Acad. of Science*, 1895, vol. xiv. p. 101.

Dr Hicks records the following zones in Pembrokeshire[1]:—

Zone of *Paradoxides Davidis* $\left.\begin{array}{c} \\ \\ \end{array}\right\}$ Menevian.

" " *Hicksii*

" " *Aurora* $\left.\begin{array}{c} \\ \\ \\ \end{array}\right\}$ Solva.

" " *Solvensis*

" " *Harknessi*

Dr Tullberg divides the *Paradoxides* beds of Scania into thirteen zones, though only a few of these are characterised by definite species of *Paradoxides*. The *Olenellus* beds have not yet been divided into zones, though this will probably be the outcome of further study[2].

The strata with *Paradoxides* are succeeded by those with the *Olenus* fauna, characterised by the genus *Olenus* and a large number of allied genera or sub-genera as some prefer to term them. The genus *Olenus* (*sensu stricto*) is very abundant in the lower part of the series, whilst the allied forms are more abundant in the upper beds. The genus *Paradoxides* and its associates disappeared before the deposition of these strata containing *Olenus* and its allies, and indeed the complete change in the character of the faunas in Europe is very remarkable. The *Olenus*

[1] The order here as elsewhere is *ascending*, i.e. the newest deposit is placed at the top.

[2] The *Paradoxides* fauna is described in the following works: Britain, Hicks, H. and Salter J. W., *Quart. Journ. Geol. Soc.*, vol. XXIV. p. 510, XXV. p. 51, XXVII. p. 173, and Hicks, H. and Harkness, R., *ibid.* vol. XXVII. p. 384; Scandinavia, Angelin, N. P., *Palæontologia Scandinavica;* Brögger, W. C., *Nyt Magazin for Naturvidenskaberne*, vol. XXIV., Linnarsson, G., *Sveriges Geologiska Undersökning*, Ser. C. No. 35; Bohemia, Barrande, J., *Système Silurien du centre de la Bohême;* Spain, Prado, C. de, "Sur l'existence de la faune Primordiale dans la chaîne Cantabrique suivie de la description des Fossiles par MM. de Verneuil et Barrande," *Bull. Soc. Geol. France*, 2 Series, vol. XVII. p. 516; America, Walcott, C. D., *Bull. U. S. Geol. Survey:* "The Cambrian Faunas of North America," and Matthew, G. F., *Trans. Roy. Soc. Canada*, 1882 and succeeding years.

fauna has been found in North Wales, Pembrokeshire
Warwickshire, Worcestershire, and abroad in Scandinavia
and Canada. It is interesting to note among the fossils
of the *Olenus* beds the occurrence of a graptolite which is
associated with *Olenus* in Scandinavia; this is the earliest
recorded appearance of a group which is destined to play
so important a role amongst the fossils of the succeeding
system[1]. The following zones have been detected by Dr
S. A. Tullberg amongst the *Olenus* beds of Scania:—

> Zone of *Acerocare ecorne.*
>   „   *Dictyograptus flabelliformis.*
>   „   *Cyclognathus micropygus.*
>   „   *Peltura scarabæoides.*
>   „   *Eurycare camuricorne.*
>   „   *Parabolina spinulosa.*
>   „   *Ceratopyge* sp.
>   „   *Olenus* (proper).
>   „   *Leperditia.*
>   „   *Agnostus pisiformis.*

The beds with *Dictyograptus flabelliformis* form a
wonderfully constant horizon at or near the top of the
Olenus beds. They are found in North Wales, the Border
Counties between Wales and England, France, Scandinavia, Russia and Canada.

The passage fauna of the beds which are the equivalents of the Tremadoc Slates may be spoken of as the
Ceratopyge fauna, for *Ceratopyge forficula*, a remarkable
species of trilobite, characterises it in Scandinavia, and

---

[1] For descriptions of the *Olenus* fauna consult the following:—Wales
Belt, T., *Geol. Mag.* Dec. I. vol. v. p. 5, and Salter, J. W., *Decades Geol.
Survey*, Decade II. Pl. IX. and Decade XI. Pl. VIII.; Scandinavia, Angelin
N. P., *Palæontologia Scandinavica*, and Brögger, W. C., *Die Silurische
Etagen 2 und 3 im Kristianiagebiet und auf Eker;* Canada, Matthew, G
F., "Illustrations of the Fauna of the St John Group, No. VI.," *Tran.
Roy. Soc. Canada*, 1891.

will probably be found elsewhere. *Ceratopyge* beds have been found in North and South Wales, Shropshire, Scandinavia, Bavaria and North America, and in each case the fauna is intermediate in character between that of the Cambrian and that of the Ordovician system, containing the loosely-formed trilobites of the former with the more compact ones of the latter. The genus *Bryograptus*, a many-branched graptolite, also appears to characterise this fauna[1].

The faunas of the Cambrian rocks have not been studied in sufficient detail, with reference to the physical surroundings of the organisms, to throw much light upon the conditions under which the strata were deposited, though the evidence obtained from an examination of the lithological characters of the deposits is generally corroborated by study of the organic contents.

[1] For accounts of the Tremadoc Slates Fauna in England and Wales see Ramsay, A. C., *Geology of North Wales*, Appendix; Hicks, H., *Quart. Journ. Geol. Soc.*, vol. XXIX. p. 39; Callaway, C., *ibid.* vol. XXXIII. p. 652, whilst many of the foreign fossils are noticed in Brögger's *Die Silurischen Etagen 2 und 3* and Barrande's *Faune silurienne des Environs de Hof en Bavière.*

CHAPTER XV.

THE ORDOVICIAN SYSTEM.

Classification. The Ordovician strata were originally divided into series by Sedgwick as follows:—

> Upper Bala,
> Middle Bala,
> Lower Bala,
> Arenig.

The Arenig series was at one time included by some writers with the Lower Bala under the name Llandeilo, but the word Llandeilo is now used in the sense of Sedgwick's Lower Bala. The Middle Bala is often spoken of as Caradoc, but the terms Bala and Caradoc are sometimes used interchangeably. As much confusion attaches to the use of the name Bala without explanation, the alternative titles have been largely adopted, and as the series are well defined there is no objection to their use, save that some expression is wanted equivalent to Upper Bala. The local term Ashgill shales was originally applied by Mr W. Talbot Aveline to beds of this age in Lakeland, and I have elsewhere suggested the use of this name for the whole series in that region; its use may well be extended to the series which is developed in many parts of Britain and the continent. The terms which will be used here, therefore, for the different series of the Ordovician system are the following:—

Ashgill Series (= Upper Bala)
Caradoc   „   (= Middle   „   )
Llandeilo   „   (= Lower   „   )
Arenig   „

Adopting a palæontological classification, we may speak
of the Arenig and Llandeilo beds as those containing
the *Asaphus* fauna, whilst the Caradoc and Ashgill beds
possess the *Trinucleus* fauna; this is the terminology
employed by Angelin for the equivalent strata of Sweden.
It must be noted that here the names applied are not
those of absolutely characteristic genera, as was the case
with those adopted for naming the Cambrian faunas, for
both *Asaphus* and *Trinucleus* range through the beds of
the system; but whereas *Asaphus* is most abundant in
the beds of the two lower series, *Trinucleus* occurs most
frequently in those of the two upper series.

Description of the strata. The Ordovician rocks are
found over large tracts in North and South Wales, in
the counties on the Welsh border, in Lakeland and the
outlying districts in the Southern Uplands of Scotland,
and in detached areas in Ireland. There are three main
types of deposit:—(i) the volcanic type, in which the
ordinary sediments are associated with a large amount of
contemporaneous volcanic matter, (ii) the black shale
type, with a fauna consisting largely of graptolites, and
(iii) the ordinary sedimentary type, in which we find
alternations of grits, shales, and more or less impure
limestones. We also find developments which are intermediate between any two or even all three of these types.
The first type is characteristically developed in Caernarvonshire and Merionethshire, the second in the Dumfriesshire
Uplands, and the third in the Girvan district of Ayrshire.

The variation in the thickness of these three types of deposit is shown in the accompanying sections of the Caernarvon, Merioneth, Moffat and Girvan regions (see Fig. 17).

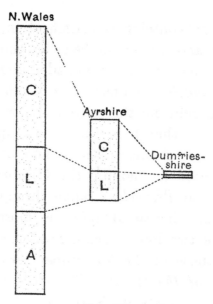

FIG. 17.

Showing the variations in the characters of the Ordovician deposits of the three principal types.

Scale 1 in. = 1000 feet.

A = Arenig.   L = Llandeilo.   C = Caradoc.

The thickness of the Arenig rocks of the Scotch areas is unknown.

The North Welsh area gives two different developments of the Ordovician strata, one of which is much less volcanic than the other. In the Merioneth-Caernarvon area, two great masses of volcanic rock form the Aran and Arenig hills of Merioneth and the Snowdonian group of Caernarvon. The former are of Arenig, the latter of Caradoc age. The Merionethshire volcanic rocks consist of a great thickness of lavas and ashes of intermediate composition (anderites), associated with sandy and muddy sediments of no great vertical depth. The Llandeilo beds of this area are chiefly of the nature of black shales, while the Caradoc series is represented by volcanic lavas and ashes of acid composition (felsites) with a few thin interbedded sediments. A calcareous ash forming the summit of Snowdon is of importance as being on the same horizon as a limestone (the Bala limestone) found in the other North Welsh area. The Ashgill series is not represented in Snowdonia.

In the other North Welsh tract, around Bala Lake, the volcanic matter is much less conspicuous. The Arenig rocks are not seen nearer than the Arenig mountains which form the western boundary of this second tract. The Llandeilo beds consist of shaley deposits with a well-marked limestone, the Llandeilo limestone, in the centre, whilst the Caradoc beds consist chiefly of muddy sediments with some thin ashes and a limestone, the Bala limestone, at the top. The Ashgill series contains a basal limestone, the Rhiwlas limestone, succeeded by shales, and another thin limestone called the Hirnant limestone at the summit.

In South Wales the Arenig beds[1] are chiefly composed of slates, and are divisible into an upper and lower group. The total thickness is about 2000 feet. The Llandeilo beds contain three series:—

Upper Llandeilo Slates 1000
Llandeilo Limestone   200
Lower Llandeilo Slates   800.

[1] A remarkable fauna, fairly well represented in Britain and exceedingly well developed on the continent, exists in the Uppermost Arenig and Lower Llandeilo beds, and it is well separated from the dominant Arenig fauna below and Llandeilo fauna above. To the beds which contain it Dr Hicks has given the name Llanvirn series.

The Caradoc beds consist of black graptolitic shales of
no great thickness, succeeded by an impure limestone on
the horizon of the Bala limestone, while the Ashgill
series like that of North Wales is separated into upper
and lower limestone stages with an intervening stage
composed of shales.

The deposits of the Welsh borderland are well
developed in Shropshire, where there is practically a
repetition of the Caernarvon-Merioneth development,
with variations in detail. The Arenig and Caradoc
volcanic rocks are not so thick as those of the Welsh
district, but are nevertheless of considerable importance[1].

In the hilly region of Cumberland, Westmorland, and
the adjoining parts of Yorkshire the succession differs from
that of any of the Welsh regions, for the great period of
volcanicity was during the formation of the Llandeilo
rocks, and there were merely sporadic outbursts in Arenig
and Caradoc times. The Arenig rocks consist of black
shales with interstratified beds of coarser sediment, and
some thin lavas and ashes of intermediate type. The
Llandeilo series is represented by a very great thickness
of volcanic rocks, varying in composition from basic to
acid lavas, with associated pyroclastic rocks. The rocks
of the Caradoc period largely consist of impure limestone
with associated argillaceous rocks, and contemporaneous
volcanic rocks of acid character. A marked unconformity
is found locally in the centre of these. The Ashgill series
consists of a basal limestone with shales above, and there
is evidence that volcanic activity had not become extinct
during the deposition of the rocks of this series.

[1] For information concerning these beds see Lapworth, C. and Watts,
W. W., "The Geology of South Shropshire," *Proc. Geol. Assoc.*, vol.
XIII. p. 297.

Passing on to Scotland, the graptolitic type is admirably shown in the southern Uplands of the neighbourhood of Moffat, Dumfriesshire. The base of the Ordovician system has not been found, but the lowest series seems to be represented by shales with a graptolite possibly of Arenig age. Above this are volcanic beds succeeded by a group of black shales known as the Moffat shales. They are only about six hundred feet in thickness, and yet represent much of the Ordovician and part of the Silurian strata as developed elsewhere. The beds belonging to the Ordovician system are divided into two series, the Glenkiln shales below and the Hartfell shales above. The former consist of intensely black muds with few fossils save graptolites, and a deposit of chert at the base which is composed of radiolaria. The graptolites of the black shales are Upper Llandeilo forms, but the thin deposit of radiolarian chert may represent the rest of the Llandeilo period and part of the Arenig period also. The Hartfell shales are also usually black graptolite shales with lighter deposits nearly barren of organic remains; they represent the Caradoc and Ashgill series and pass conformably into the deposits of Silurian age[1]. The ordinary sedimentary type of Ordovician rocks is found in Ayrshire, though a few thin graptolitic seams are

---

[1] The Moffat beds are described in a paper by Prof. Lapworth entitled "The Moffat Series" in the *Quarterly Journal of the Geological Society*, vol. xxxiv. p. 239. This paper, which is a masterpiece of detailed work, has furnished a clue to many problems. Few students will be able to follow the numerous details, and for general information concerning the beds they are recommended to read another paper by the same author "On the Ballantrae Rocks of South Scotland," *Geol. Mag.* Dec. iii. vol. vi. p. 20. An account of the radiolarian cherts by Dr G. J. Hinde will be found in the *Annals and Magazine of Natural History* for July, 1890, p. 40.

intercalated with the conglomerates and shelly sands, clays and limestones of the region, which is therefore peculiarly valuable as affording a means of comparison of the shelly type with the graptolitic type of Ordovician deposits. The Arenig series consists of black shales with graptolites, and these rocks are succeeded by a volcanic group which is probably of Llandeilo age. Above these volcanic beds, as in Dumfriesshire, we find three great divisions, two of which are of Ordovician, the third of Silurian age. The Ordovician divisions are respectively termed the Barr series, which is the equivalent of the Glenkiln shales, and the Ardmillan series above, equivalent to the Hartfell shales [1].

It is interesting to find that in the north of Ireland the rocks generally coincide in characters with those which are found along the same line of strike in Great Britain; thus, the Girvan type appears in Londonderry, Tyrone and Fermanagh, the Moffat type in County Down, and the Lake District type in the counties of Dublin and Kildare.

On the continent the volcanic material which plays so important a part in the constitution of the Ordovician accumulations of Britain is practically absent, and the strata are largely composed of accumulations of shale and limestone with occasional coarser deposits. In Scandinavia, the Arenig beds consist of limestones with a few shales, the Llandeilo deposits are largely calcareous, those of Caradoc age are partly calcareous and towards the top usually argillaceous, while the equivalents of the British

---

[1] See Lapworth, C., "The Girvan Succession," *Quart. Journ. Geol. Soc.*, vol. xxxviii. p. 537, and also the paper on the Ballantrae Rocks referred to in the preceding footnote. The latter paper should be carefully read by all students of the stratigraphy of the Lower Palæozoic Rocks.

Ashgill series are calcareous at the base and argillaceous at the summit. In Russia the calcareous matter preponderates over the argillaceous material.

Ordovician strata are also found in Belgium, France, Bohemia, and other places, and are largely composed of mechanical sediments of varying degrees of fineness mixed occasionally with some calcareous matter.

The variation in the characters of the Ordovician strata of Britain points to accumulation in a fairly deep sea, usually at some distance from the land, but dotted over with volcanoes which often rose above the water, causing the addition of much volcanic material to the ordinary sediments, and the existence of minor unconformities at different horizons along their flanks. As these unconformities are not always associated with volcanic material it is obvious that uplifts must have occurred occasionally during the deposition of the rocks; one important uplift is indicated by the occurrence of an unconformity in the Arenig rocks of Wales, while another is seen amongst the Caradoc rocks of the Welsh borders. On the whole, however, the period was one of slow subsidence, the deposition of material generally keeping pace with this subsidence, and accordingly there is a great uniformity of characters amongst the strata over wide areas. The probable continuation through the Ordovician period of the tract of land over the present site of the N. Atlantic ocean which as we have reason to suppose existed during Cambrian times, is indicated by similar changes of lithological character amongst the strata when traced from Britain eastward to Russia in both Cambrian and Ordovician times, and the continuance of these conditions over the American area is also indicated by study of the variations amongst the American Ordovician deposits.

The Ordovician Faunas. The Ordovician period has justly been termed the Period of Graptolites, which are the dominant forms of the time, and continue in abundance throughout the period. The abundance of graptolites in black shales associated with few other organisms has often been noted. It appears to be due to a large extent to the slow accumulation of the graptolitic deposits allowing an abundance of these creatures to be showered upon the ocean floor, after death, for the evidence derived from detailed examination of their structure points to their existence as floating organisms. The tests of other creatures largely calcareous may well have been dissolved before reaching the sea-floor. In support of the view that these black shales are abysmal deposits may be noted the singular persistence of their lithological characters over wide areas, their replacement by much greater thicknesses of normal sediments along the ancient coast-lines, the frequent occurrence together of blind trilobites with those having abnormally large eyes when these creatures are associated with graptolites in the black shales, and lastly the interstratification of the black shales with radiolarian cherts similar to the modern abysmal radiolarian oozes. If this be so, we ought to find graptolites in marine deposits of all kinds, and indeed they are found there, though largely masked by the mass of sediment and the hosts of other included fossils, so that their discovery is rendered much more difficult than when they occur in the black shales,—a state of things which is familiar in the case of other pelagic organisms as *Globigerinæ*, radiolaria, and pteropods, whose tests abound in the abysmal deposits and are comparatively rare in those of terrigenous origin[1].

[1] The importance of the graptolites as indices of the geological age will

The characters of the Ordovician trilobites have already been noticed. These organisms are abundant, and occur in sediments of all kinds. Of other groups, the significance of the radiolaria has been referred to above. Corals occasionally form reef-like masses of limestone as in the limestones of the Caradoc epoch; the echinoderms are well represented, cystids being locally abundant; of the crustacea, many remains of tests of phyllocarida have been recorded; the brachiopods are very abundant, and of the mollusca, lamellibranchs, gastropods and cephalopods all occur with frequency though none of these groups is very prevalent. Certain forms have been referred to pteropods though with doubt, and other shells seem to be referable to the heteropods. The existence of vertebrates during Ordovician times is not, in the opinion of many geologists, proved, though remains of fishes have been recorded from the Ordovician strata of North America; but it is desirable that more evidence of this occurrence should be given [1].

The distribution of the Ordovician faunas like that of the sediments points to the prevalence of open ocean conditions over wide areas during the period, with occasioual approaches to land, which was often of a volcanic nature. Around this land clustered the ordinary invertebrates, building up coral-reefs and shell-banks, whilst away in the open oceans the graptolites floated, almost alone, and sank to the ocean floor after death.

be seen by perusal of Prof. Lapworth's paper "On the Geological Distri- bution of the Rhabdophora," *Ann. and Mag. Nat. Hist.*, Ser. 5, vol. III. (1897).

[1] Walcott, C. D., "Preliminary Notes on the Discovery of a Vertebrate Fauna in Silurian (Ordovician) Strata," *Bulletin Geol. Soc. America*, vol. III. p. 153.

CHAPTER XVI.

THE SILURIAN SYSTEM AND THE CHANGES WHICH OCCURRED IN BRITAIN AT THE CLOSE OF SILURIAN TIMES.

Classification. The Silurian system was originally divided by its founder, Sir R. I. Murchison, into three series, as follows :—

Ludlow Series

Wenlock „

Llandovery „

The term May Hill, proposed by Sedgwick, is sometimes used as synonymous with Llandovery. This classification omits a somewhat important set of beds intercalated between those of the Llandovery and Wenlock series known as the Tarannon shales, and in Britain if we were to classify afresh, it would be more convenient to include some of the beds formerly referred to the Ludlow in the Wenlock. I shall, however, adopt the old and well-established classification, adding the term Tarannon to Llandovery, and speaking of the Llandovery-Tarannon series. The nature of the two classifications is shown in the following table:

Stages.	Old Classification.	New Classification.	Palæontological Classification.
1 Upper Ludlow		Downtonian	
2 Aymestry Limestone	Ludlow		
3 Lower Ludlow			Fauna with *Encrinurus*
4 Wenlock Limestone			
5 Wenlock Shale	Wenlock	Salopian	
6 Woolhope Limestone			
7 Tarannon Shales			Fauna with *Harpes*
8 Upper Llandovery	Llandovery	Valentian	
9 Lower Llandovery			

Description of the strata. Lithologically the Silurian deposits of Britain form a continuation of those of the Ordovician period, with a local interruption due to the elevation of portions of Wales and the Welsh borders at the close of Ordovician times. Elsewhere we find a predominance of shales passing into grits at the top of the system, the change indicating the incoming of the shallow-water phase before the commencement of the second continental period. Particular stress is laid upon the predominant shaley character of the beds, for, on account of the richness and variety of the faunas of the calcareous rocks, greater attention is naturally paid to them in geological works, and the student may get a false idea of their relative importance. An attempt is made below (Fig. 18) to give a general idea of the variations in lithological

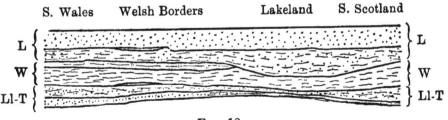

FIG. 18.

L = Ludlow.   W = Wenlock.   Ll-T = Llandovery-Tarannon.

characters of the Silurian rocks in different parts of Britain.

The Silurian strata are mostly found in the same localities as those which furnish exposures of the rocks of Ordovician age.

The development in the typical Silurian region of the Welsh borders is characterised by the abundance of calcareous matter which is found there as compared with that which exists in the other British localities.

The Llandovery strata are sandy, often conglomeratic, with a fair amount of calcareous matter in places. The arenaceous nature is undoubtedly due to the proximity of land caused by local upheaval at the end of Ordovician times, and the Upper Llandovery rocks sometimes rest unconformably on the Lower ones, at other times on Ordovician, Cambrian, or even Precambrian rocks. The Tarannon shales are light green shales with intercalated grits. The Wenlock series consists of a group of shales separating a lower, very inconstant, earthy limestone from an upper, more constant, thicker and purer limestone. The latter, the Wenlock limestone, is composed of frag-ments and perfect specimens of various fossils, and the fragmentary nature of many of the shells indicates the occurrence of wave-action and probable formation in shallow water, in some places against coral-reefs.

The Lower Ludlow beds consist of sandy shales; they are separated from the Upper Ludlow beds by an impure limestone, the Aymestry limestone. The Upper Ludlow beds consist mainly of grits and flags, often coloured red towards the summit.

In North Wales the Llandovery beds occasionally present the shelly arenaceous types of deposit as near Llangollen, at other times as near Conway, Corwen, and in Anglesey, the graptolitic shale type. They also rest unconformably upon the Ordovician rocks in this area. The Tarannon shales resemble those of the border county. The Wenlock series consists essentially of shales, while the Ludlow development differs from that of the borders in its greater thickness and the absence of any calcareous band in the centre. In Central Wales the graptolitic type of the Llandovery-Tarannon series is found, but the graptolite-bearing shales of the Llandovery epoch are thin beds occurring between grits and flags no doubt deposited in shallow water, and this division of the series is of very great thickness.

In South Wales the Silurian rocks are very similar to those of the Welsh borders, save that the calcareous deposits are fewer and thinner.

The Lake District Silurian strata generally resemble those of North Wales. The Llandovery-Tarannon rocks are of the graptolite-shale type, intercalated with fine grits in the case of the beds of Tarannon age. The Wenlock beds consist of shales, and the Ludlow beds of gritty shales beneath, and massive flags and grits at the summit. These Ludlow beds are here of great thickness (certainly not less than 7000 feet) and were obviously accumulated for the most part in shallow water.

The Llandovery-Tarannon rocks of Southern Scotland show the two types which prevailed in the Moffat and Girvan areas in later Ordovician times. The Llandovery beds of Moffat are known as the Birkhill shales, and are very thin. The representatives of the Tarannon shales, however, the Gala beds, consist mainly of grits, and attain a great thickness. In the Girvan area, the Llandovery beds are of the shelly type. Here as at Moffat and in the Lake District there is perfect conformity between the beds of Ordovician and those of Silurian age, and accordingly it is instructive to note the completeness of the palæontological break, especially in the Moffat district. The higher Silurian beds of Southern Scotland present a general resemblance to those of North Wales and the Lake District[1].

On the European continent we find indications of conditions similar to those which prevailed during the Ordovician period; the strata become much thinner and more calcareous in Scandinavia, and still thinner in the Baltic provinces of Russia, where they consist very largely of calcareous matter. In central Europe the greater abundance of calcareous matter, compared with that which is found in the Ordovician strata of that region, points to a change in physical conditions which became still more marked after Silurian times.

In North America, the succession is very similar to that of Britain, the calcareous development of the Silurian rocks being found around Niagara, but towards the close of Silurian times the shallow-water phase became marked in places by the deposition of chemical precipitates which indicate the separation of a portion of the late Silurian ocean from the main mass during the period of formation of these abnormal deposits.

The conditions of Silurian times, until the advent of the shallow-water phase, recall those of Ordovician times and point to a wide expanse of ocean at some distance

---

[1] For descriptions of the Silurian beds of the typical region see Lapworth and Watts, *Proc. Geol. Assoc.*, vol. XIII. p. 297, those of Wales are described by Lake and Groom, *Quart. Journ. Geol. Soc.*, vol. XLIX. p. 426, and Lake, *ibid.* vol. LI. p. 9. A description of those of Lakeland will be found in the Memoir of the Geological Survey "The Geology of the Country around Kendal, etc." while the Scotch Rocks are described in Lapworth's papers on Moffat and Girvan.

from the land, though the earliest deposits become arenaceous where they were deposited against an old land surface formed by the elevation of the Welsh Ordovician rocks, which were denuded to supply this material. One marked difference existed between the physical conditions of our area during Ordovician and Silurian times, for the volcanic activity which was rife during Ordovician times almost ceased during Silurian times, except in the region now occupied by the extreme south-west of Ireland, and accordingly volcanic material does not appreciably contribute to the formation of the Silurian deposits. The shallowness of the sea-floor at times is marked by the occurrence of masses of reef-building corals in the limestones, and these probably indicate the prevalence of a fairly warm climate, an inference supported by the nature of the Gastropod fauna of Gothland, as noticed in Chap. IX.

The shallow-water phase commences fairly simultaneously over the whole area at the beginning of the deposition of the Lower Ludlow rocks, and becomes more marked in the Upper Ludlow rocks, being most noticeable at their extreme summit, when a change occurred which will be considered at the conclusion of this chapter.

The Silurian Faunas[1]. The Silurian period has been termed the period of Crinoids, and this group of creatures certainly contained a great variety of very remarkable forms, which are specially numerous in the Wenlock Limestone of the Welsh borders, Gothland, and North America, but many of the rocks of the system display few traces of these organisms. The trilobites and grapto-

---

[1] For an account of the Silurian faunas the student may consult Sir R. I. Murchison's *Silurian System* or the shorter *Siluria* and Lapworth's paper on the Geological Distribution of the Rhabdophora.

lites still contribute largely to the fauna, the latter
becoming very scarce at the summit of the system
though a few specimens have been detected in the rock
of the succeeding system. The trilobites belong to few
genera though these are mostly more highly organised
than those of the Ordovician period. The genus *Harpe*
has been taken as fairly characteristic of the lower part
of the system in Sweden, and it occurs there abundantly
in places in Britain, whilst *Encrinurus* is more abundant
in the upper series, but both of these genera range from
higher Ordovician beds into the Devonian. Mention has
already been made of the corals. Brachiopods are very
abundant, and Mollusca appear with considerable fre
quency. The appearance of true insects is of importance
cockroaches have been recorded from Silurian rocks and
a number of other insects have lately been recorded from
Canada[1]. Eurypterids occur in considerable abundance
in the higher parts of the system, as do also the remains
of fish.

The close of Silurian times ushered in the second
continental period in Britain when a large part of our
area and the adjoining areas to the north and north-east
were uplifted to form land, which in the case of our
area was interpenetrated by watery tracts, whose exact
nature is still a subject of dispute. Accordingly the
deposits which were formed during this period are local
and in some cases abnormal, but they will be considered

---

[1] See an article by Dr G. F. Matthew, "Description of an extinct
Palæozoic Insect and a review of the Fauna with which it occurs.
Bulletin xv. *of the Natural History Society of New Brunswick.* The
Silurian Rocks of the Little River Group of St John, New Brunswick
have yielded species of land snails, two doubtful saw-bugs, several
arachnids, and myriopods, two insects of the order Thysanura (Spring
tails), and eight Palæodictyoptera.

in the next chapter. Simultaneously with the formation of these deposits, uplift of the sea-floor converted wider and wider areas into land, and this land underwent considerable denudation, so that the tops of the anticlines were worn away. The general trend of the anticlines was east-north-east and west-south-west, and accordingly a series of mountain chains possessed that direction, for the epeirogenic movements were accompanied by orogenic ones. Between the regions of uplifts were depressions in which sediments accumulated. The principal axes of uplift in our area range through the North of Scotland towards Scandinavia, across the Southern Scotch Uplands to the North of Ireland, through the Lake District and through Wales. As the result of lateral pressure, a cleavage structure was impressed on many of the Lower Palæozoic rocks, the strike of the rocks extended in the direction of the ridges and depressions, and the rocks as a whole became considerably compacted and hardened, thus producing one of the most important portions of the framework of our island, for although the ancient mountain chains were largely denuded during their elevation, and their stumps were afterwards covered by later deposits, upon the removal of these, the ancient stumps were once more exposed as fairly rigid masses which do not yield greatly to denuding influences, and accordingly stand out as the most important upland regions of Britain at the present day.

It is interesting to notice, as an illustration of the now well established fact that successive earth movements often occur in the same direction, that the axes of the folds produced during this second continental (Devonian) period, run parallel with the lines separating tracts of different lithological characters. It has been seen that the Ordovician and Silurian rocks of the Southern Uplands continue into Ireland, and that the beds of similar characters run in belts having a general east-north-east and west-south-west trend, which accordingly must have been the direction of the coast-line parallel to which they were deposited, and as that coast-line was due to uplift, the movement which produced it would naturally produce foldings with east-north-east and west-south-west trend. This is one of many cases where the lines separating belts of rock having different lithological characters run parallel to axial lines of folds which have been produced in the rocks at a later period.

As the result of the existence of land over parts of north-west Europe in Devonian times, it is comparatively rare to find a passage from normal Silurian rocks into normal Devonian ones; there is often an unconformity above the Silurian strata. As we proceed southwards towards central Europe, where the epeirogenic and orogenic movements died out, this is not the case, and we get complete conformity between marine sediments of the Silurian and Devonian periods.

CHAPTER XVII.

THE DEVONIAN SYSTEM.

Classification. As a result of the movements which were briefly described in the last chapter, two types of Devonian deposit are found in the British Isles, and are called respectively the Devon type and the Old Red Sandstone type. The latter rocks, formerly divided into three divisions, are now separated into two only, the upper and lower Old Red Sandstone, and the exact relation of these to the different subdivisions of the rocks of Devon type remains to be settled. The Devon type itself has given rise to much difference of opinion, two local classifications have been applied, one for the rocks of North Devon and another for those of South Devon. The classification which has been most generally adopted is as follows :—

	N. Devon.	S. Devon[1].
Upper Devonian (Clymenian)	Pilton Beds Cucullæa (Marwood) Beds Pickwell Down Sandstone	Entomis Slates Goniatite Limestones and Slates Massive Limestones

[1] An account of the South Devon rocks by Mr Ussher will be found in the *Quart. Journ. Geol. Soc.*, vol. XLVI. p. 487; from it the above classification of the rocks of S. Devon is taken.

	N. Devon.		S. Devon.
Middle Devonian (Eifelian)	{ Morte Slates, Ilfracombe Beds	{	Middle Devonian Limestones, Ashprington Volcanic Series, Eifelian Slates and Shaly Limestones
Lower Devonian (Coblenzian)	{ Hangman Grits, Lynton Slates, Foreland Grits	{	Lower Devonian Slates, Lincombe and Warberry Grits and Meadfoot Sands

The division into Lower Middle and Upper Devonian is generally adopted, though the alternative titles given to these divisions are not always used with the same signification, and the distribution of the different local stages given in the above classifications is usually adopted in the main, though a detailed comparison of the Devonian beds of North and South Devon is still attended with difficulty.

More than once an attempt has been made to prove that the apparent succession of the North Devon rocks, which is that given in the above table, is not the true one, and of recent years Dr Hicks has obtained a number of fossils from the Morte Slates which had hitherto yielded none, and he believes that these fossils indicate that the Morte Slates are on a lower horizon than the beds on which they rest. Whatever be the ultimate verdict, we can, at any rate, say that the " Devonian Question," as it is termed, is not settled[1].

Description of the Strata. The general variations in

[1] See Hicks, H., "On the Morte Slates and Associated Beds in North Devon and West Somerset," *Quart. Journ. Geol. Soc.*, vols. LII. p. 254; LIII. p. 438.

the lithological characters of the deposits of Devonian age will be seen from the accompanying figure which represents the deposits of Britain as they occurred from north to south before they had been affected by subsequent earth-movements (Fig. 19). The conventional signs which are used are similar to those which have been used in other parts of this work, and will save description of the section.

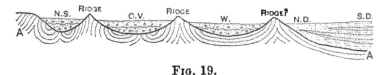

FIG. 19.

A. Lower Palæozoic and Precambrian Rocks.

N.S. North of Scotland ⎫
C.V. Central valley of ditto ⎬ Old Red Sandstone Type.
W. Wales ⎭

N.D. North Devon ⎫
S.D. South Devon ⎬ Devon Type.

The ridges separate different deposits of Devonian rocks, which were possibly deposited in isolated areas, though there was probably connexion between them at any rate at times.

The Old Red Sandstone type consists to a large extent, as the name implies, of sandstones which are coloured red by a deposit of peroxide of iron around the sand grains. They are separable into a lower and upper division with an unconformity often occurring between them. The lower Old Red passes down in places into the Silurian rocks with perfect conformity, and the upper Old Red similarly passes up into the Carboniferous strata. The existence of pebble beds at different horizons is a noteworthy feature. They are frequently found at or near the base of the two divisions. The sandstones of the lower division are often accompanied by flagstones, while the red sandstones of the upper division usually have deposits of yellow and brown sandstone intercalated between them. Inconstant beds of limestone, known as cornstones, are found in both divisions, and Prof. Sollas has shown that some of these, at any rate, are true mechanical deposits, formed by the destruction of pre-existing strata of limestone and the deposition of the resulting fragments from a state of suspension. In Scotland a great thickness of volcanic material of various kinds is associated with the two divisions. For the sake of simplicity this is omitted from Fig. 19[1]. It is not known how far normal sediments are associated with the Old Red Sandstone type of deposit. The existence of some in South Wales is suggested by evidence supplied by the late Mr J. W. Salter.

The Devon type, as will be seen in the figure, consists of rocks which are to a great extent of normal character. We find in Devonshire alternations of sandstones, shales and limestones, but even here, red sandstones, which are comparable with those of the Old Red type occur in diminished amount: the Foreland Grits and Pickwell Down Sandstones are both coloured red, and are like the sandstones formed further north. The recognition of this fact induces one to believe that the contrast between the two types of rock which are found at a short distance from one another on opposite sides of the Bristol Channel is not so marked as one is sometimes led to suppose.

The rocks of North Devon differ from those of South Devon chiefly owing to the amount of calcareous sediment found in the two areas, for limestones occur in South

---

[1] For an account of these and all other British volcanic rocks the reader is referred to Sir A. Geikie's work on *The Ancient Volcanoes of Great Britain.* Macmillan and Co., 1897.

Devon to a great extent, and in North Devon there is a comparative poverty of this kind of sediment. Here, again, the apparent difference is possibly greater than the real one. The North Devon limestones have in places been stretched out after their formation and thus rendered thinner, and the highly-cleaved limestones are occasionally mistaken for shales, while in South Devon there is evidence of thickening of the limestones by folding subsequently to their deposition. Allowing for these changes, however, there is still a marked diminution in the amount of coarse mechanical sediments and increase in the quantity of calcareous matter as one passes from North to South Devon, and this prepares one for the condition of things met with on parts of the continent, where the mechanical sediments become finer and thinner on the whole as one travels southward, until, when we reach the Bohemian area, the Devonian rocks are found to be largely composed of calcareous sediments.

It is interesting to find that in North America the two types of Devonian strata recur, and present characters generally similar to those which they possess upon this side of the Atlantic.

Passing now to a consideration of the conditions under which the Devonian rocks were deposited, we may examine the bearing of the character of the strata as a whole, and then proceed to more detailed consideration of the nature and conditions of deposits of the two types.

The gradual increase in calcareous matter and dying out of mechanical sediments as one travels southward points to recession from land in that direction, and we have already seen that the epeirogenic and orogenic movements of this continental period elevated the Silurian sea-floor in the north, and gave rise to a Northern

Continent, while oceanic conditions continued further
South, and allowed the accumulation of sediments lying
conformably upon those of Silurian age, and giving indications of the prevalence of physical conditions during
Devonian times which were in the main similar to those
of the preceding Silurian period.

In the shallow waters adjoining the land of the
Northern Continent the Old Red Sandstones were laid
down, and the exact conditions under which they were
accumulated is a matter of some interest. The late Sir
Andrew Ramsay gave reasons for supposing that many
red deposits were accumulated in the waters of inland
lakes, which underwent rapid evaporation, and his views
have been applied, with much corroborative evidence by
Sir A. Geikie, to account for the red sandstones of Devonian age, which he believes to have been accumulated
in a series of inland lakes, though others hold a different
opinion, and consider that the Old Red Sandstone waters
had a direct connexion with those of the open ocean;
the question is too intricate to be discussed at length
here. Besides the difference of physical characters of
the two types of strata, the difference in the nature of
their included organisms is significant. The ordinary
invertebrates, as corals, crinoids, brachiopods and molluscs
are extremely rare in the Old Red Sandstone, which
contains remarkable remains of Agnatha fishes and eurypterids, and although these are also found associated with
a true marine fauna in Russia, Germany and Bohemia,
the rarity or apparent absence of the ordinary marine
invertebrates, though only negative evidence, which is
proverbially dangerous, must be regarded.

The North Devon rocks are sediments which might
well be accumulated on the shores of a continent, while

those of South Devon, with their abundant coral reefs, and other organic limestones were no doubt deposited in a clearer sea, at a greater distance from the land, and the clear water deposits of Germany and still more of Bohemia, were accumulated in the open ocean. It is interesting to note in these Bohemian deposits abundance of shells of a Pteropod *Styliola* which has been proved by Prof H. A. Nicholson to form masses of limestone in the Devonian system of Canada. The modern distribution of the Pteropoda suggests the open ocean character of the deposits which contain them even so far back as Devonian times, though one cannot conclude that these deposits are really analogous to the so-called Pteropod ooze of modern seas which, as a matter of fact, is largely composed of foraminiferal tests with a considerable percentage of pteropod shells.

The Devonian flora and faunas. The plant remains in the Lower Palæozoic rocks are few in number. Some undoubted terrestrial plants have been discovered, but the prevalent flora of lower Palæozoic times, so far as yet known, was one consisting of Algae. In Devonian times we begin to meet with a number of Cryptogams of higher type, allied to those which form the dominant flora of the succeeding period. The fauna is in many ways remarkable. The Devonian period has been termed the period of ganoid fishes, and the remarkable remains, so graphically described by the late Hugh Miller, are indeed peculiarly characteristic of Devonian times, but they are largely though by no means exclusively entombed in rocks of the Old Red Sandstone type[1]. The Devon type of rock contains a great abundance and variety of the problematical group, the Stromatoporoids, which contribute extensively

[1] For an account of these see A. S. Woodward's *Vertebrate Palæontology*.

to the formation of many of the limestones, and although these organisms are not by any means confined to Devonian strata, their abundance and variety therein might lead one to speak of the period as that of Stromatoporoids. The remains of corals are very abundant in the limestones, and, as already stated, frequently give rise to true reef-masses. The graptolites, as remarked in the previous chapter, disappear in the rocks of the Devonian period, and as only one or two fragments have been found, we may assert that the group was practically extinct at the end of Silurian times, though species of one genus, *Monograptus*, lingered for a short time in greatly diminished quantity. The trilobites which played so important a part amongst the faunas of Lower Palæozoic times still occur fairly abundantly amongst the rocks of the Devonian system, and there is a very interesting point to be noticed in connexion with them. They seem to have become practically extinct in the succeeding Carboniferous period, where few genera are found, and the decadence of the group began in Devonian times. In these circumstances it is interesting to note the tendency displayed by the creatures to possess spiny coverings. It is true that *Acidaspis*, the most spinose of all trilobites, is abundant in Ordovician and Silurian strata, and that other spinose trilobites are found there, but the peculiarity of the Devonian trilobites is, that genera which were previously smooth, or rarely possessing one or few spines, are found represented by extremely spinose species in these beds,—the spines being developed from all parts of the test, sometimes as a fringe to head or tail, sometimes as prominent projections from glabella and neck segment, and frequently in rows down the body segments. Besides *Acidaspis*, we find spinose species of *Phacops, Homalonotus, Cyphaspis, Bronteus* and *Encrinurus* in Devonian strata, and the occurrence of these forms is so frequent and world-wide, that one might perhaps infer with confidence that an unknown fauna containing many spiny trilobites was of Devonian age.

The abundance of Eurypterids has been previously noted. Occurring as they do in Silurian rocks, they are far more abundant in those of Devonian age, and are found indifferently in sediments of Old Red and Devon types. Of air breathers, several insects have been found in the strata of different parts of the world.

The ordinary marine faunas are otherwise intermediate in character between those of the Silurian and Carboniferous periods, but there are several characteristic Devonian genera, and no one who is acquainted with the peculiarity of the Devonian fauna would deny to the Devonian strata the right to rank as a separate system, containing a fauna as well marked in its way as that of the Silurian system below or that of the Carboniferous above. Special stress is laid upon this point because it has been suggested that the Devonian system should be abolished, and its strata either divided between the Silurian and Carboniferous systems or referred exclusively to the latter system[1].

[1] The literature of the fauna of the Devonian rocks is a rich one. For an account of the Devonian rocks of Britain, the reader may consult the Monograph of the Devonian Fossils of the South of England by Rev. G. F. Whidborne, which is now appearing in the series of Monographs of the Palæontographical Society, and in the publications of the same Society he will find a Monograph of the Eurypterids from the pen of Dr Henry Woodward. The richest Devonian fauna is undoubtedly that of the Bohemian area, for the work of Dr E. Kayser has conclusively proved that the stages *F*, *G* and *H* of that basin, formerly referred to the Silurian, are of Devonian age, and an excellent idea of the richness of the Devonian fauna may be obtained by studying the descriptions of the fossils from those stages which have appeared and are appearing in Barrande's classic work.

CHAPTER XVIII.

THE CARBONIFEROUS SYSTEM.

The Classification. The British rocks of the Carboniferous system have been classified according to their lithological characters, but as the classification has been altered from time to time, we may use that which seems most acceptable to the majority of British geologists at the present day. According to this, the beds are grouped as below :—

Upper Carboniferous	Coal Measures	Ardwick Stage Pennant Stage Gannister Stage
	Millstone Grit	
Lower Carboniferous	Carboniferous (Mountain) Limestone Series.	

The Lower Carboniferous beds have been further subdivided into :—

Yoredale Series or Upper Limestone Shales,
Mountain Limestone,
Lower Limestone Shales, with Sandstones and Conglomerates,

but as these lithological types are found to be very variable when traced laterally for comparatively short distances, it is found more satisfactory to use the terms in a purely lithological sense rather than with chronological significance.

The somewhat abnormal development of the higher portions of the Carboniferous rocks of Britain renders the local classification only partially applicable in other regions, and as our knowledge progresses, a palæontological classification will probably be adopted. This has already been done with the more purely open-water sediments of Russia and Eastern Asia, where the development of the beds is more normal. There the rocks are classified as under :—

Upper Carboniferous or Gshellian,
Middle Carboniferous or Moscovian,
Lower Carboniferous,

and as this classification has already been found to be applicable over rather wide areas, it is almost certain that, as in the case of the rocks of other systems, it will prove more serviceable than one which is mainly (though not quite exclusively) based upon vertical variation of lithological characters, especially as the Carboniferous rocks over large tracts in North America possess faunas which are similar to those which have been discovered in Russia, Eastern Asia and North Africa.

Description of the strata. The variations in the lithological characters and fossil contents of the British Carboniferous strata when traced from north to south have been so frequently described, and utilised as a means of illustrating the indications as to local variations in physical conditions which are supplied by those strata, that little need be said upon the subject. The restoration of the physical geography of Carboniferous times over the British area will be found in a chapter by the late Professor Green in the work upon *Coal* by various professors at the Yorkshire College of Science and also in Prof. Hull's *Physical History of the British Isles.* Some

M.

modifications must be made in these restorations as th
result of recent research, the principal being caused t
discoveries amongst the Carboniferous rocks of Devoi
shire.

Taking the strata in vertical succession, we find ev
dence of the occurrence of a complete marine period (th
second great marine period) between the second and thii
continental periods.   The first shallow-water phase over
great portion of the British Isles is marked by thi
terrigenous sediments, indicating that the period was
brief one; it was followed by the deep-water phas
probably of some length, lasting through the greater pai
of the remainder of Lower Carboniferous times; while th
concluding shallow-water phase was lengthy as compare
with that of the beginning of the period, and is marke
by the accumulation of the great thickness of deposil
belonging to the Millstone Grit and Coal Measure
There is no doubt, however, that in some parts of th
British area minor changes produced local terrestri;
conditions during the period, and accordingly we find th;
the deepest water deposits of the system in Britain ai
succeeded by an unconformable junction with the sed
ments of the upper portion of the system.

The general change in the lithological characters (
the beds of the Lower Carboniferous division when trace
from south to north is shewn in the following diagra'
(Fig. 20).

It will be seen that the land and open sea areas wei
in the respective positions which they occupied durii
Devonian times, but that as the result of greater sui
mergence, with which the accumulation of sedimei
did not keep pace, the shallow-water marine deposi
of Devonian age are in Devon replaced by open-s(

deposits[1], while shallow-water marine deposits further north replace the anomalous deposits which were found there during the Devonian period.

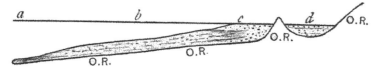

FIG. 20.

a. Radiolarian cherts of Devon.
b. Mountain limestone of Central England.
c. Mechanical sediments of Northern England.
d. Freshwater deposits of Southern Scotland.
O.R. Older rocks.

Owing to the accumulation of thick masses of sediment, the Lower Carboniferous sea of the north of England appears to have been largely silted up, and although the organic deposits of the south are so thin that they did not render the sea shallow in that region, the general level of the Lower Carboniferous floor of the south was also uplifted, and actually converted into land, as the result of the upward movement which took place in Devonshire and tracts of France; and owing to silting up in the north, and elevation in the south, a general plane surface was produced over very extensive areas, not only in Britain but upon the Continent, upon which the peculiar deposits and accumulations of Upper Carboniferous times were laid down, sometimes in shallow water, sometimes upon the land, and often under conditions which cannot at present be determined with accuracy. That the deposits of the Millstone Grit and Coal Measure

[1] The Radiolarian Cherts of the Lower Carboniferous rocks of Devon, and the associated sediments, together with the unconformity between these and the Upper Carboniferous beds are described by Messrs Hinde and Fox, *Quart. Journ. Geol. Soc.*, vol. LI. p. 609.

epochs were to a large extent laid down in water is admitted by all, and in the case of many of the deposits of the Millstone Grit, and some thin deposits of the Coal Measures, it is equally clear that the water area was part of an expanse of ocean, for we find marine fossils, as corals, crinoids, and cephalopods, in these beds. Associated with them in the Coal Measures are other beds in which the ordinary Carboniferous genera of marine invertebrates are absent, and their place is taken by shells which bear much resemblance to the modern fresh-water mussel, and it has been maintained with good reason that as the ordinary marine forms are rarely or never mixed with those resembling recent fresh-water shells, the latter are truly fresh-water[1]. If this be so, many of the mechanically formed sediments of the Coal Measures were of fresh-water origin, laid down in shallow lagoon-like expanses, probably shut off from the main ocean by a narrow portion of intervening land, which was occasionally destroyed, thus permitting incursions of salt-water when some of the ordinary marine invertebrates of the period obtained a temporary footing in the area.

There is not only a difference of opinion as to the mode of accumulation of many of the mechanical sediments of the Coal Measures, but also as to that of the coal-seams which accompanied them. Two different theories have been put forward to account for these coal-seams, which are usually spoken of as the drift theory and the growth-in-place theory. According to the former, in its extreme

---

[1] For further information upon this subject the student should consult the Introduction to a Monograph on *Carbonicola*, *Anthracomya* and *Naiadites* (the shells in question) by Dr Wheelton Hind, being one of the Monographs of the Palæontographical Society.

application, coal is an aqueous deposit formed by the settlement of drifted masses of vegetation upon the floor of a water-tract, while those who push the growth-in-place theory to its extreme limits maintain that coal is the result of growth of vegetation upon the actual site where the coal is now found. Much apparently conflicting evidence has been advanced by the advocates of the two hypotheses, and special cases of coal-formation have been appealed to by each in support of their views; thus the existence of coal composed largely of bodies which resemble the spores of modern lycopods,—objects of so resinous a nature that they float on the surface until they are decomposed,—is cited by the upholders of the growth-in-place theory, while the supporters of the other hypothesis can point with equal force to the occurrence of the finely divided carbonaceous mud containing remains of fishes which gives rise to cannel coal in some places. One of the main assertions in support of the growth-in-place theory was that of the supposed universality of 'underclays' or old surface soils beneath all coal-seams, but though these are common, they are far from universal. It is impossible to do justice in small compass to this question of coal-formation, but it may be pointed out that much of the difference of opinion can be understood if it be remembered that the term 'coal' is rather a popular term which has been admitted into scientific terminology, and therefore used somewhat loosely, than a strictly scientific term applied to a definite substance, and accordingly, just as at the present day we find carbonaceous substances growing in one place on land to form peat, in other places on a tract sometimes dry and sometimes submerged, to form the carbonaceous deposits of the cypress-swamps, and once more accumulated beneath the shallows of a sea as a sediment to form the carbonaceous muds of the ocean margins where the mangroves grow, so the diverse substances which are included under the general term coal may have accumulated in one place on land, in another beneath water, and in a third on an area alternately dry and submerged. This is not a question of great importance; the important point is that accumulations of vegetation on a fairly large scale are found at the present day on plains, for even if they grow on mountain regions, the deposits are readily denuded before they are covered up, and also it must be noted that a moist climate is necessary for the growth of much vegetation. The conclusion that the accumulations of coaly matter were formed on plains is borne out by their great horizontal extent as compared with their thickness, and it is now generally agreed that the coal vegetation which is found in the normal coal-measures was essentially a swamp vegetation.

An attempt has been made to prove that an upland vegetation of very different character existed contemporaneously with it, but reasons will be given in the sequel for concluding that this supposed upland Carboniferous flora is everywhere of later date.

The later shallow-water phase of Carboniferous times, as already stated, was unusually long, it was also very wide-spread, and appears to have been accompanied over wide areas by humid conditions during its continuance, and accordingly the marsh conditions which existed during Upper Carboniferous times were probably on a larger scale than that of similar conditions before or after. Special stress is laid upon this fact, as it is a good illustration of the view which seems to be gaining ground, that every period possessed peculiar conditions never to be repeated, which must have left their impress upon the character of the sediments.

Though the conditions above described were widespread, they were naturally not universal, and accordingly in many parts of the world, as previously stated, we find true marine deposits of Upper Carboniferous times, though even these were sometimes replaced during part of the epoch, by conditions which were favourable for the formation of coal-seams in those places. Interruption in the continuance of a humid temperate climate over the regions of North-West Europe is also suggested by the discovery of deposits which are maintained to be of glacial origin amongst the Coal Measures of France [1].

The Floras and Faunas. The flora of the Carboniferous rock is so noteworthy that the period has been termed the Period of Cryptogams; the remains of ferns, horsetails, and clubmosses predominate, and many of the forms reached a gigantic size. Though the floras of the various stages are marked by a general resemblance, there are differences which enable the palæobotanist to ascertain the stratigraphical position of the beds by reference to the included plant remains, and a considerable number of successive floras have been described [2]. The invertebrate fauna does not differ on the whole very greatly from that of Devonian times, though the trilobites are now becoming rare, and the mollusca assume a more prominent position as compared with the brachiopods. Corals occur in abundance in the calcareous deposits of the period, and frequently give rise to sheets of reef-

[1] For an account of the numerous volcanic products see Sir A. Geikie's work on "The Ancient Volcanoes of Great Britain."

[2] Consult Kidston, R., "On the Various Divisions of the Carboniferous Rocks as determined by their Fossil Flora," *Proc. Roy. Phys. Soc. Edin.*, vol. xii. p. 183.

formation, but the foraminifera and crinoidea certainly play the principal part as limestone-producers, and the influence of the latter in giving rise to great masses of limestone which are frequently used for ornamental purposes is too well known to need more than passing reference. The air-breathers have also been detected in greater abundance, though they are rare, when we consider the comparatively favourable conditions for their preservation presented by the Coal Measure rocks. Myriopods, arachnids, insects and pulmoniferous gastropods have however been found with tolerable frequency. The danger of arguing from imperfect data is well illustrated by the great addition to our knowledge of the insect-fauna of these times due to the exploration of the beds of one small coal-field, that of Commentry in France, of which the insects have been described by M. C. Brongniart. The vertebrates are represented by a considerable variety of fishes, and less abundant though tolerably numerous remains of Amphibia, which occur in the Carboniferous rocks of the North of England, Ireland, France, North America and elsewhere.

The existence of definite zones of organisms in the case of the Carboniferous rocks has been denied, and it appears to be considered by some that the Carboniferous rocks were accumulated so rapidly as compared with rocks of some other systems that the fauna remained very similar throughout. It is very doubtful if this was so. In the case of other systems, the division into zones has only been accomplished by means of more detailed researches than those which have been conducted amongst the Carboniferous rocks of Britain: again, the occurrence of successive floras suggests that there may have been a similar succession amongst the faunas, and finally we find that zonal division has been carried on to some extent amongst the Carboniferous strata of other regions. The following classification of the Russian type of sediment may prove useful, as an indication of the possibility of more detailed separation of our own beds :—

Gshellian (with *Fusulina* and *Archimedipora*)	Beds with *Spirifera fascigera, Spiriferina Saranae,* &c. Beds with *Producta cora, P. uralica, Camarophoria crumena,* &c. Beds with *Syringopora parallela* and *Spirifera striata.*
Moscovian	Stage of *Spirifera mosquensis.*
Lower Carboniferous	Stage of *Spirifera Kleini.* Coals, Sandstones and Shales with *Noegerathia tenuistriata* and *Producta gigantea.* Stage of *Producta mesoloba.*

The marine fauna of the Upper Carboniferous beds, which is so poorly represented in Britain, but is well developed in Spain, Russia, Asia and North America, is largely characterised by the abundance of foraminifers of the genus *Fusulina* and *Fusulinella* and of bryozoa of the genus *Archimedipora*. It is very desirable that the truly marine fauna of the *Spirorbis* limestone and other marine bands of the British Coal Measures should be carefully studied to see if they present any close relationship with that of the Gshellian beds[1].

[1] A good idea of the general characters of the Carboniferous fauna of Britain will be obtained from an examination of Professor Phillips' *Geology of Yorkshire*, Part I., and Mr (now Sir F.) McCoy's *Carboniferous Fossils of Ireland*, while the nature of the European fauna is well illustrated in Prof. de Koninck's well-known work *Description des animaux fossiles qui se trouvent dans le terrain carbonifère de Belgique*. For an account of the characters of the marine fauna of the Upper Carboniferous rocks the reader should consult the work on Geology and Palæontology published by the Geological Survey of the State of Illinois in 1866.

CHAPTER XIX.

THE CHANGES WHICH OCCURRED DURING THE THIRD CONTINENTAL PERIOD IN BRITAIN; AND THE FOREIGN FERMO-CARBONIFEROUS ROCKS.

AT the close of Carboniferous times a marked change took place in the nature of the earth-movements. The prevalent depression which occurred over the British and adjoining regions during Carboniferous times was replaced by upward movement, accompanied by orogenic folds, which once more brought on continental conditions and developed a series of mountain ranges. The change is marked even at the close of Carboniferous times by the abnormal red sandstones of the uppermost part of the Carboniferous system which are found around White-haven in Cumberland and Rotherham in Yorkshire, as the Whitehaven Sandstone and Rotherham Red Rock. These movements continued through Permian and Triassic times, and it is to them and to the climatic conditions of the periods, that the anomalous nature of the Fermo-Triassic deposits is largely due, as will be shewn in the succeeding chapters. At present it is our purpose to call attention to the effect of these movements upon the sediments which had been deposited previously to their occurrence.

Over the British area, two different systems of orogenic movement can be detected, producing folds of which the axes run approximately at right angles to one another. One of these, of which the Pennine system is the best representative in Britain, caused the production of elevations having axes in a general north and south direction, and we may therefore speak of it as the Pennine system of movement, while the other, which gave rise to folds running in an east and west direction, is well represented in the Mendip Hills, and may be therefore termed the Mendip system, though it is more widely known as the Hercynian system, as, on the Continent, the rocks which are greatly affected by it form the foundations of the region occupied by the ancient Hercynian forest.

The effects of these systems were in the main similar; they resulted in the uplift of parallel belts of country to form hill-ranges with intervening lowlands, but when studied in detail the movements are seen to be of a different character. The Pennine system of movements was of a type which is familiar to the geologists as developed in the Great Basin Region of the western territories of North America, and produced what is spoken of as Basin-Range structure. The movements were of the nature of direct uplift, causing fracture, only accompanied by folding in a minor degree, and accordingly the hills are composed of terraced scarps, with one gently sloping side, and one steep scarp-side, the latter on the upthrow side of the fault, as seen in fig. 21.

In the Mendip system, the folds were of the Alpine type, which is a familiar product of lateral pressure, consisting essentially of overfolds, though these are often complicated by reversed faults.

Of the Pennine system, the Pennine Chain itself fur-

nishes the most noteworthy example in Britain, but we
have indications of other folds of this system, such as that
which runs from the Lake District to the Ayrshire coast,
which is partly concealed as the result of other movements,
and a still more marked one, in the rocks of the Malvern
Hills.

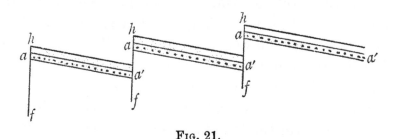

FIG. 21.

a a'. One stratum displaced by faults *ff*.   *h.* **Hills.**

The Mendip system is well shewn in the Mendip Hills,
but the remains of a still more important anticline are
seen in South Devon and Cornwall, separated from the
Mendip Hills by the great syncline of Devon. Another
parallel anticline runs from Lancashire to Yorkshire at
right angles to the Pennine Chain and separates the coal-
field of Cumberland and that of Northumberland and
Durham, from those of South Lancashire, and Yorkshire,
Notts, and Derbyshire.

On the European continent the Ural Chain is the
most important uplift of the system of which the Pennine
Chain forms a minor representative, while the Hercynian
system has caused the compression and stiffening of many
of the Carboniferous and earlier rocks which now rise to
the surface in many parts of central Europe.

The extensive continental area which was the result
of these uplifts not only determined the formation of
abnormal deposits, but allowed the occurrence of a long

period of time subsequently to the close of the Carboniferous period, of which few deposits now exposed in Europe
are representative, and we must accordingly seek other
regions in order to find typical representatives of this
Permo-Carboniferous period, of which the strata developed
in the Salt Range of India have been most carefully
worked, especially by Dr Waagen, though marine sediments of the period are known elsewhere, as in Spitsbergen,
the Ural Mountains, China and Australasia; and a group
of somewhat anomalous sediments of this age in parts of
India, Australia and South America is of peculiar interest,
on account of the insight as to the climatic conditions of
the times which it affords.

The Permo-Carboniferous Rocks. In the Salt Range
of the North-West of India an interesting series of sand-
stones alternating with limestones rests unconformably
upon lower rocks. The sandstones are known as the
Speckled Sandstones, while the limestones are termed the
Productus Limestones. The Lower and Middle Speckled
Sandstones are succeeded by the Lower *Productus* Limestone which is separated from the Lower division of the
Middle *Productus* Limestone by the Upper Speckled
Sandstone; these are all of the Fermo-Carboniferous
period, while the upper part of the Middle *Productus*
Limestone and the Upper *Productus* Limestone belongs
to the Permian period. The fossils, largely invertebrates,
are intermediate in character between those of Carboniferous and Permian ages. Similar fossils are found in the
marine Fermo-Carboniferous beds of the other areas which
have been named above. The Lower Speckled Sandstone
is of interest on account of the occurrence of boulder-beds
within it, and this division of the sandstone has been
correlated with the lowest (Talchir) stage of the Fermo-

Carboniferous strata of other parts of India, while the other Speckled Sandstones and those divisions of *Productus* Limestone which are referred to the Fermo-Carboniferous are correlated with the higher divisions of other parts.

Special mention is made of the Talchir division, on account of the occurrence therein of boulder beds which have long been known, and whose glacial origin was inferred by Dr W. T. Blanford forty years ago. The accumulations shew signs of having been deposited in water, but the existence of large subangular, sometimes striated boulders therein, which must have come from distant sources, and the occasional occurrence of striated rock surfaces on the strata upon which the Talchir beds repose unconformably points to ice-action; this would not be so very remarkable if it were an isolated case, though sufficiently so, from the comparative nearness of the region to the equator; but researches conducted in different parts of the southern hemisphere have brought to light similar, and sometimes even more striking evidences of glacial action in widely distinct regions[1]. In Australia they have been found in New South Wales, Victoria, South Australia, East Australia and Tasmania; the Dwyka boulder-conglomerates of South Africa and certain deposits of similar character discovered by Prof. Derby in Southern Brazil have been referred to the same period, and their glacial origin has also been inferred. This wide-spread distribution of deposits which are generally contemporaneous, of which the glacial origin may now be taken as established, is extremely remarkable,

[1] The reader will find an excellent account of the Permo-Carboniferous glacial deposits in a paper by Prof. Edgworth David, entitled "Evidences of Glacial Action in Australia in Fermo-Carboniferous Time" (*Quart. Journ. Geol. Soc.* Vol. LII. p. 289). In this paper other glacial beds besides those of Australia are noticed.

and must be taken into careful consideration by those
who put forward theories framed to account for former
climatic changes.

The Flora and Fauna. The flora of the Fermo-Carboniferous beds has caused as much discussion as the
question concerning the origin of the boulder-deposits.
In the southern hemisphere, the Fermo-Carboniferous
rocks of those countries which have yielded boulder-beds
also contain remains of a flora which is now known as
the *Glossopteris* flora, from the prevailing genus, which
is associated with other genera, such as *Gangamopteris*.
These fossils appear to be ferns, though their modern allies
have not been indicated with certainty; associated with
them are rare cycads and conifers. The *Glossopteris* flora
is markedly contrasted with the Coal-Measure flora of the
northern hemisphere with its giant lycopods. Moreover
Glossopteris appears in the northern hemisphere in rocks
of later date than the Fermo-Carboniferous period. It
has been suggested that the *Glossopteris* flora originated
in a continent in the southern hemisphere, on which the
boulder beds were also formed in isolated water areas, and
that some of the forms migrated northwards. To this
continent the name Gondwanaland has been applied by
Prof. Suess, from the *Gondwana* series of the Fermo-Carboniferous rocks of India, in which the *Glossopteris* flora is
found, and it has also been maintained that the southern
Glossopteris flora was contemporaneous with the northern
flora of ordinary Coal-measure type, though whether this
was so to any extent remains to be proved, for the beds
containing the *Glossopteris* flora are distinctly newer than
any which have furnished a typical northern Coal-measure
flora. In any case, the change of floras between Coal
Measure and Fermo-Carboniferous times is very marked,

and when taken in connexion with the wide-spread glacial deposits, is one of the most striking phenomena displayed by the rocks of the stratified column[1].

The fauna has already been noticed. It consists of brachiopods, some of which are of peculiar genera. The general similarity of the faunas in regions so remote as Spitsbergen, the Ural Mountains, India, and New South Wales, indicates an extensive sea during the period. It can hardly be supposed that the fauna of Fermo-Carboniferous times has been completely described, for the fossils of one or two areas only have been made known to us with any degree of fulness, and when the Fermo-Carboniferous and marine Permian faunas are as well known as those of Triassic times (and the latter have only been fully described very recently) there is no doubt that the important break which was at one time supposed to exist between Palæozoic and Mesozoic faunas will be filled in satisfactorily[2].

[1] For an account of the Glossopteris flora and its geological relations, consult Seward, A. C., *Science Progress*, January, 1897, p. 178.

[2] The Fermo-Carboniferous beds are described in Dr Blanford's *Geology of India*, second edition (edited by Mr R. D. Oldham), and figures of some of the important fossils given therein. For fuller information the reader should refer to Waagen's account of the Salt Range Fossils and Feistmantel's description of the plants in the *Memoirs of the Geological Survey of India*.

CHAPTER XX.

THE PERMIAN SYSTEM.

Classification. It has already been observed that as the result of the Pennine and Mendip systems of earth-movement, the Carboniferous rocks of Britain are succeeded by a marked unconformity, and that the rocks of the succeeding Permian and Triassic systems of Britain shew an abnormal development. The principal areas where Permian rocks are found are on either side of the Pennine Chain in the North of England, but sporadic exposures of rocks of this age are found in some of the Midland and Southern counties. The Permian rocks have been well studied in Germany, and the German names are sometimes adopted in Britain, and the following comparison will prove useful :—

Britain.	Germany.	
Magnesian Limestone	Magnesian Limestone	} Zechstein.
Marl Slate	Kupferschiefer	
Lower Permian Sandstones	Rothliegende.	

The term Zechstein has been applied in a somewhat different sense by different writers, but the one given in the table appears to find most favour.

In a region which was essentially continental, considerable variations in the lithological characters of the rocks may be expected, when the strata are traced laterally, but we nevertheless find that the differences are not so great as was formerly supposed to be the case when certain red sandstones lying above recognised Permian strata in the district on the west side of the Pennine Chain towards its northern extremity were also referred to the Permian; these sandstones (the St Bees Sand-stones) are now generally admitted to be of Triassic age, and comparison between the rocks on opposite sides of the Pennine Chain is much simplified, as seen below.

West side.	East side.
Thin Magnesian Limestones and Marls	Magnesian Limestone
Hilton Shales	Marl Slate
Penrith Sandstone and Brockrams	Lower Permian Sandstones.

Description of the Strata. On the east side of the Pennine Chain, the Lower Permian sandstone is an in-constant deposit often consisting of yellow false-bedded arenaceous strata. The Marl Slate is an argillaceous shale, often containing bituminous matter, and yielding several fish-remains and some plants; it is usually only a few feet in thickness. The Magnesian Limestone is typically developed in Durham as a yellow or greyish limestone containing a variable percentage of carbonate of magnesia; when traced southward, it alters its charac-ters, becoming mixed with mechanical deposits, and some chemical precipitates in places, so that at Mansfield it appears as a red sandstone with grains cemented by a mixture of carbonates of lime and magnesia; and, like the rest of the Permian strata, it has disappeared when we reach Nottingham. In addition to the southward thin-ning of the Permian beds of this area, there is some evidence of their disappearance in a westerly direction, though, as the present strike of the beds is nearly north and south, the indications of this are less convincing.

On the east side of the Pennine Chain, the main difference observable is the relative thickness of the major divisions. The Lower Permian sandstones have thickened out considerably, while the reputed representatives of the Magnesian Limestone are thin. The Penrith sandstone is of considerable interest. It contains in places, as near Appleby, thick deposits of breccia consisting of angular fragments chiefly composed of Carboniferous Limestone, which in many cases have undergone subsequent dolomitisation, embedded in a matrix of red sandstone. This breccia is known as brockram. Many beds of the Penrith sandstone are composed of crystalline grains of sand, due to deposition of silica in crystalline continuity with the quartz of the original grain after the formation of the deposit; of more significance, for our present purpose, is the presence of other accumulations of the sand, in which the individual grains often approach the form of spheres, thus resembling the ' millet-seed' sands of modern desert regions. The Hilton shales are grey sandy shales, with plant remains, and above them are variable deposits including thin magnesian limestones which have yielded no fossils.

The isolated Permian deposits of the midland and southern counties of England consist of red marls and sandstones with occasional breccias, and in the absence of fossils, their exact position in the Permian series is still unknown.

The German Permian rocks resemble those of Britain, especially as seen in Durham, in many particulars, and give indications of formation under physical and climatic conditions generally similar to those which were then prevalent in the British area. At Stassfurt, in Germany, the less soluble constituents of ocean water are accompanied by a great variety of salts :—chlorides, sulphates and borates; and the very soluble salts of potassium and magnesium known as the Abraum salts are found in abundance as well as the less soluble salts of sodium and calcium. The occurrence of these very soluble salts is so infrequent on a large scale among the rocks of the Geological Column, and the matter is one of so great theoretical import, that it is necessary to take special note of their presence in the Permian strata.

The frequent existence of chemical deposits in the Permian Rocks of N.W. Europe, the formation of red sandstones, and the dolomitisation of limestone beds and fragments of pre-existing limestones point to inland seas of a Caspian character, while the evaporation necessary for the formation of the precipitates also indicates a fairly warm temperature. The presence of millet-seed sands, in very lenticular patches, suggesting former sand-dunes, and the occurrence in places of breccias (like some parts of the brockram) almost devoid of matrix, piled up against pre-existing cliffs, recalling screes of modern times, give almost certain evidence of the occurrence of land tracts most probably of desert character, during part of the period of accumulation of the materials of the Permian rocks. The fossil evidence supports this view, and geologists are mostly agreed that the Permian rocks of north-west Europe were accumulated in an area of desert character, occupied in part by inland seas, though there is much difference of opinion as to the extent of these seas, some geologists holding that a number of isolated sheets of water were necessary to produce the distribution and character of the accumulations. It is still a vexed question with British geologists how far the Pennine ridge stood up as land during the period, but leaving this and other minor considerations out of account, it may be noted that the similarity of deposits in the different areas, whether we examine the order of succession, the lithological characters or the included fossils, suggests communication between the water tracts of different regions, though this communication need not have been more than a series of straits, or comparatively narrow belts of water[1].

The extensive development of Permian and Triassic rocks with terrestrial characters in the southern hemisphere also, and the absence of newer deposits in many places, suggests that the land areas of these times in that hemisphere have largely remained such ever since, in which case, the Fermo-Triassic series of movements produced a marked direct effect upon our present continental areas, and at any rate produced an indirect one upon the British land tracts.

The presence of anomalous deposits of Permian age over wide areas need not be surprising, but it would be indeed remarkable if no ordinary marine type of Permian rocks was known, and the researches of recent years have proved that this type is extensively developed, in Eastern Europe, Asia, and North America, where Permian rocks consisting of limestones, with a greater or less admixture of mechanical deposits, occur in some abundance. The

---

[1] It should be mentioned that some writers have inferred the evidence of glacial conditions over parts of the British area, on account of the resemblance of some of the Permian breccias to recent glacial deposits. The question is still *sub judice*. It is not necessarily opposed to the existence of desert conditions, if the mountains were sufficiently high, for the Wahsatch regions adjoining the Basin Region of N. America have been glaciated.

studies of Waagen and others in India have given us the farthest insight into the nature of these beds. Below is a general classification taken from Waagen's work :—

Salt Range.	Germany.
Base of Trias Unfossiliferous Shale and Sandstone Top Beds of Upper *Productus* Limestone	Passage Beds into Trias
Cephalopoda Beds of Upper *Productus* Limestone	Gypsum Beds
Middle Division of Upper *Productus* Limestone Lower Division of Upper *Productus* Limestone	Zechstein (in restricted sense)
Upper Division of Middle *Productus* Limestone	Weissliegende and Kupferschiefer
Middle Division of Middle *Productus* Limestone	Rothliegende.

It will be seen that in the Salt Range there is a complete passage from the Fermo-Carboniferous strata into the Trias, and the detailed work which has been carried out by Waagen and others amongst the rocks of the Salt Range must make this, for the present at all events, the type area for the marine development of the strata of Fermo-Carboniferous and Permian ages.

The Permian flora and fauna. The Permian flora presents some difficulties. The flora of the Zechstein consists largely of ferns and conifers, but that of the Rothliegende of Germany has been compared with that of the Carboniferous, and if a true Permian flora of the northern hemisphere has many forms of Carboniferous affinities, the presence of the Glossopteris flora in Permo-Carboniferous rocks of more southerly regions seems to imply its origin there and *slow* migration northwards. It must be noted, however, that the Rothliegende has been divided by some geologists into an upper and lower division, of which the lower is actually referred to the Carboniferous system. All that can be now said is, that our knowledge of the floras of Fermo-Carboniferous and Permian times is still incomplete, and that the difficulties will no doubt be cleared up as the result of further work.

The invertebrate fauna of the north-west European Permian deposits is chiefly noticeable on account of the paucity of species, though individuals are often abundant. The shells are also sometimes stunted and occasionally distorted. These characters bear out the supposition that the aqueous deposits were laid down in inland seas of Caspian character and not in the open ocean. Polyzoa, brachiopods, and lamellibranchs predominate, but other groups are found. The vertebrates consist of forms of fish, amphibia and reptiles, and the Permian rocks are the earliest strata in which the remains of true Reptilia are known to occur with certainty. The Reptiles belong to the orders Anomodontia (Theromora) and Rhynchocephalia, of which the former is exclusively Permian and Triassic, while the latter is abundant in the strata of those periods, but is represented at the present day by the genus *Sphenodon* of New Zealand. The Amphibia belong to the order Labyrinthodontia which ranges from Carboniferous to Lower Jurassic, but the members of the order are most abundant in Permian and Triassic strata, and these periods may be spoken of as the Periods of Labyrinthodonts.

A few words must be said of the fauna of the truly marine Permian beds. It is much richer than that of the abnormal deposits of north-western Europe, and its study is important as furnishing another link between Palæozoic and Mesozoic life. Many Palæozoic genera pass up into the Permian rocks, and, as will be ultimately seen, several occur in those of the Triassic system, and one or two even in the basal Jurassic strata, though Mesozoic forms predominate in the Lower Jurassic Rocks, and there is a fairly equal admixture of forms usually considered as Palæozoic and of those generally regarded as Mesozoic in Triassic rocks, while the Palæozoic forms still predominate over the Mesozoic in the Permian strata. Along with these characteristic Palæozoic genera, it is interesting to find representatives of more than one genus of the tribe of Ammonites, which is to take so prominent a place in the fauna of the Mesozoic rocks, amongst the true marine Permian sediments of India and other areas. The announcement of the contemporaneity of ammonites with fossils regarded as exclusively palæozoic was received with considerable doubt, but this contemporaneity is now clearly established, and need not be regarded as in any way anomalous.

With the deposition of the Permian rocks, Palæozoic time comes to an end, but as already remarked there is no marked and sudden change to characterise it. Had our classification been originally founded on study of the Indian Rocks instead of those of Britain, and similar terms adopted, the line of demarcation between Palæozoic and Mesozoic rocks would probably have been drawn below the Fermo-Carboniferous deposits, and if it had been based on study of other areas, perhaps elsewhere. The palæontological break is purely local, and it is of the utmost importance that it should be recognised as such, and that it should not be considered that division into

Palæozoic and Mesozoic implies some great and widespread change which occurred between the times covered by the deposits of each of these great divisions[1].

[1] The Permian Fossils of Britain are described by Professor King in the Monographs of the Palæontographical Society (the Brachiopods by Dr Davidson in the Monographs of the same Society). For a general account of the marine type the student may consult the second edition of Dr Blanford's *Geology of India.*

CHAPTER XXI.

THE TRIASSIC SYSTEM.

Classification. The term Triassic has been applied to these rocks on account of the threefold division into which those of Germany naturally fall. These three divisions are :—

Keuper,
Muschelkalk
Bunter ;

but above the Keuper beds we find a group of deposits of some importance, which shew affinities with both Triassic and Jurassic rocks, which may be looked upon as true passage beds, though they are generally placed in the Triassic System. They are known as Rhætic or locally in Britain as Penarth Beds. The Muschelkalk is usually considered to be unrepresented in Britain, and accordingly the British deposits may be, and are usually grouped as under :—

Rhætic or Penarth beds

Keuper
{ Keuper Marls
{ Keuper Sandstones

[Muschelkalk] absent

Bunter
{ Upper Red and Mottled Sandstones
{ Bunter Pebble Beds
{ Lower Red and Mottled Sandstones.

The threefold grouping has been applied more or less universally, but when used outside the north-west European area, it loses its significance, as the conditions which enable one to differentiate the rocks of the three divisions were naturally only prevalent over a limited area.

Description of the strata. The British Triassic rocks possess a certain sameness as regards their general characters, consisting mainly of mechanical sediments coloured red by peroxide of iron, with occasional chemical precipitates of rock-salt and gypsum. They have a wider distribution over Britain than have the Permian rocks, and the lithological characters of the different subdivisions do not as a rule vary to a remarkable degree when traced laterally. The differences in detail in the characters of the various deposits are noteworthy, and an explanation of the exact origin of some of these abnormal deposits which will satisfy everyone is not yet forthcoming. Leaving the details out of consideration for the moment, and looking at the general aspect of the deposits, the prevalence of conditions generally similar to those which existed over the British Isles in the preceding Permian period is decidedly indicated by the nature of the strata, though the continental conditions appear to have been more widely established over our area, as shewn by the general absence of any calcareous deposits resembling the Magnesian Limestone. We find chemical precipitates, millet-seed sandstones, and scree-like breccias in the British Triassic rocks as well as in those of Permian age, and the paucity of a marine invertebrate fauna in the Triassic rocks of Britain is even more apparent than in the Permian strata. It is only at the extreme close of the Triassic period, during the deposition of the rocks which are admitted on all hands to be of Rhætic age, that we note the incoming of those marine conditions over our area, which prevailed so extensively, with few local exceptions, during the remainder of the Mesozoic and the early part of Tertiary times; the Rhætic beds, in fact, mark the commencement of the third marine period. Referring to the strata in further detail, we may proceed to consider the character of the different subdivisions in the order of their formation, commencing as usual with the oldest. The Bunter deposits rest in places upon those of Permian age with an unconformity at the junction, but as these unconformities occur frequently among the British Triassic rocks, it is doubtful whether this unconformity marks more than very local change of physical conditions. The lower and upper divisions of the Bunter sandstone consist of false-bedded red and variegated sandstones, and there is no great difficulty in explaining their formation in desert areas with tracts of water, but the great change which marks the appearance and disappearance of the middle division, the Bunter pebble beds, requires some explanation, for the contrast between the lithological characters of the rocks of this division and those of the rocks appertaining to the preceding and succeeding division is very marked. The matrix differs, but the main difference is the abundance of pebbles, mostly of fairly uniform size, well rounded, and largely consisting of liver-coloured quartzite. Much difference of opinion exists as to the exact origin of these pebble beds, and the source of the pebbles, but without entering into this vexed question, it may be remarked that the agency of rivers has been somewhat generally invoked to account for their transport, and the conditions during their ac- cumulation need not have been very different from those which are now found in northern India where the torrential rivers of the south side of the Himalayan chains debouch upon the plain, and spread an abundant deposit of well-worn pebbles over the finer silts which were previously laid down thereon.

The junction of the Bunter and Keuper beds requires a short notice. It is usually if not always an unconformable one in Britain, and it is generally assumed that the absence of the Muschelkalk of the Continent is due to the presence of land undergoing denudation in Britain during the time when the Muschelkalk was elsewhere deposited, though it is quite possible that the Muschelkalk epoch is represented in Britain not only by the time which elapsed when the unconformity was being impressed on the rocks, but also during the true deposition of the upper part of the Bunter beds, or the lower part of the Keuper, or both.

The Keuper sandstones and marls contain a great development of chemical deposits, of millet-seed sands, and of many other features pointing to desert conditions, such as sun-cracks, tracks of animals impressed upon a rapidly drying surface, and pseudomorphs of mud after rock salt in the form of cubes and hopper-crystals; furthermore we find the scree-like breccias at different horizons of the Keuper beds where they abut against the old Mendip ridge composed largely of mountain-limestone which furnished the fragments, as was the case with the brockrams abutting against the Pennine ridge. It must be noted that the chemical precipitates of Triassic age consist of the less soluble substances dissolved in ocean water, namely, gypsum and rock salt, whilst the more deliquescent potash and magnesia salts are not represented in Britain.

Turning to these continental beds, we get evidence of

a general approach to open sea conditions as we pass away
from Britain in a south-easterly direction as roughly shewn
in the following diagram (fig. 22), where *B* represents the
Bunter beds, *M* the Muschelkalk, and *K* the Keuper.

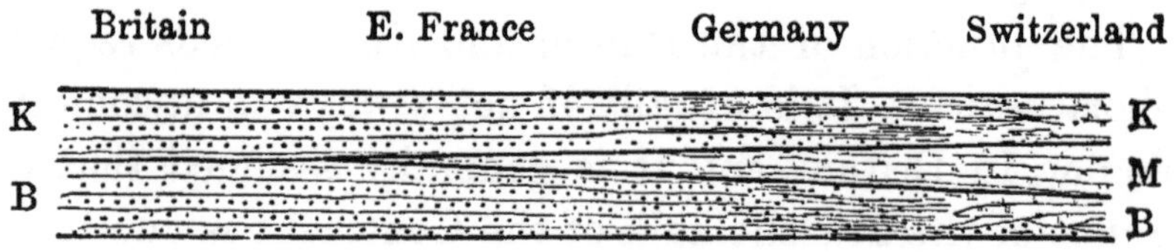

FIG. 22.

It will be seen that the mechanical sediments gradually die out and become replaced by calcareous material
as one passes from Britain towards Switzerland; the
Muschelkalk is very thin in the east of France and
thickens out in Germany, while in Switzerland Keuper,
Muschelkalk and Bunter are alike largely represented by
calcareous deposits, and the mechanical deposits are
chiefly argillaceous, the only important sandstone being
situated at the extreme base of the Bunter series.

The marine development of the Triassic system is naturally the one which is most widely spread, though full
appreciation of its importance has only taken place as
the result of researches in distant climes of recent years.
It is found in southern Europe, in Spitsbergen, in considerable tracts of Asia, including India, and along the
Pacific coast region of North America, and everywhere
possesses much the same characters.

It will be seen from the above remarks that the
physical conditions which prevailed in the continental
area of Triassic times which is now partly occupied by the
British Isles are most closely represented by those of the
desert regions of central Asia, hemmed in by the mountain

ranges which intercept the vapour-laden winds of the oceans, and cause them to precipitate the great bulk of their vapour on the seaward slopes of the mountains, so that they blow over the deserts as dry winds, causing the fall of any large amount of rain to be a rare though by no means unknown event in the desert regions.

Flora and Fauna of the Period. The Triassic flora is essentially similar to that of the higher Permian strata, though many of the genera are different.

The invertebrate fauna of the British deposits is, as might be expected, very poor until the beds of the Rhætic series are reached. In the beds below the Rhætics, the principal invertebrate remains are the tests of the crustaeean genus *Estheria,* though a few obscure lamellibranch shells have been recorded. The vertebrate fauna is of great interest. A number of fishes have been found, the most remarkable of which is the genus *Ceratodus,* occurring in the Rhætic beds of Britain and lower Triassic strata of foreign countries. It is closely related to the Barramunda of the Queensland rivers belonging to the order Dipnoi. As in the Permian strata, abundance of Labyrinthodont amphibians have been discovered, and the reptiles belong to the orders Anomodontia and Rhynchocephalia. In the Rhætic beds of Britain and in still lower Triassic beds abroad the orders Ichthopterygia and Sauropterygia (represented by *Ichthyosaurus* and *Plesiosaurus*) are found.

The Triassic rocks also yield the earliest known mammals, the best known, *Microlestes,* occurring in the Triassic rocks of Britain and the Continent. These mammals are now placed in a subclass Metatheria of the order Monotremata.

The marine invertebrate fauna of the normal Triassic

rocks presents some points of considerable interest. As
already remarked, the fauna may be looked upon as a
passage fauna between that of Palæozoic and that of
Mesozoic times, the number of Palæozoic forms which
pass into the Trias being approximately comparable with
those which appear here and range upwards into higher
Mesozoic strata. This may be well seen by examining
the table given in Chapter XXI. of the Second Edition of
Sir Charles Lyell's *Student's Elements of Geology*, in which
three columns shew the genera of Mollusca common to
older rocks, those characteristic of the Trias, and those
common to newer rocks. Amongst the first are *Orthoceras*,
Bactrites, Loxonema, Murchisonia, and *Euomphalus,* in the
second column are *Ceratites, Halobia (Daonella), Ko-
ninckina,* and *Myophoria,* and in the third, Ammonites,
Cerithium, Opis, Plicatula and *Thecidium*[1].

The Ammonites are largely utilised in the case of the
Mesozoic strata for separation of these strata into zones,
each zone being characterised by some species of Ammonite, and the researches of Mojsisovics have proved
that this zonal subdivision, long adopted for Jurassic
rocks, is also applicable to those of Triassic age[2]. He
gives the following table of the classification of the
Triassic rocks of the Mediterranean Province, which is
reproduced, as it is founded upon Palæontological evidence, and will probably be widely adopted.

---

[1] It has been seen that some of the Ammonites appear earlier, namely,
in Permian strata. *Myophoria* is extremely abundant in the Trias, but
ranges into newer strata.

[2] von Mojsisovics, Dr E., "Faunistische Ergebnisse aus der Unter-
suchung der Ammoneen-faunen der Mediterranen Trias." *Abhandl. der
k. k. Geologisch. Reichsanstalt.* VI. Band 2 Abtheilung. Vienna, 1893.

Series		Zonal Divisions
RHAETIC		1. Zone of *Avicula Contorta*
JUVAVIC	Upper Juvavic	2. Zone of *Sirenites Argonautae*
		3. Zone of *Pinnacoceras Metternichi*
	Middle Juvavic	4. Zone of *Cyrtopleurites bicrenatus*
		5. Zone of *Cladiscites ruber*
	Lower Juvavic	6. Zone of *Sagenites Giebeli*
CARNIC	Upper Carnic	7. Zone of *Tropites subbullatus*
	Middle Carnic	8. Zone of *Trachyceras Aonoides*
	Lower Carnic	9. Zone of *Trachyceras Aon.*
NORIC	Upper Noric	10. Zone of *Protrachyceras Archelaus*
	Lower Noric	11. Zone of *Protrachyceras Curionii*
MUSCHELKALK	Upper Muschelkalk	12. Zone of *Ceratites trinodosus*
	Lower Muschelkalk	13. Zone of *Ceratites binodosus*
BUNTSANDSTEIN	Werfener Schichten	14. Zone of *Tirolites Cassianus*

CHAPTER XXII.

THE JURASSIC SYSTEM.

THE Jurassic rocks were formerly separated on account of differences of lithological character into Oolites and Lias, but it was apparent that the Oolites were more important than the Lias, and a fourfold division was made into:—

Upper or Portland Oolites }
Middle or Oxford Oolites } = Malm
Lower or Bath Oolites   = Dogger
    Lias.

The Lias strata have also been spoken of as the Black Jura, the Lower Oolites and part of the Oxford Oolites as Brown Jura, and the rest of the Oxford Oolites with the Portland Oolites as White Jura.

As the outcome of a detailed study of the faunas of the Jurassic rocks, a further subdivision has been made, partly based upon the original British series, but the divisions are defined with greater accuracy, so that they are applicable over wider areas. They are as follows:—

Upper Oolites { Purbeckian
              { Portlandian
              { Kimmeridgian

Middle Oolites	{ Corallian Oxfordian Callovian
Lower Oolites	{ Bathonian Bajocian
Lias	{ Toarcian Liassian Sinemurian.

Many of these series have been still further subdivided into smaller stages, and the whole differentiated into a number of zones characterised by different forms of Ammonites. Dr E. von Mojsisovics gives thirty-two Ammonite zones, of which fourteen occur in the Lias, eight in the Lower Oolites, six in the Middle Oolites, and four in the Upper Oolites.

Characters of the strata. The whole of the Jurassic rocks and also those of Lower Cretaceous age may be regarded as having been deposited during the first shallow water phase of the third marine period, but this shallow water phase is represented by strata which are varied owing to numerous marine changes resulting in the production of land at times, and estuarine conditions, shallow water, marine conditions, and somewhat deeper sea conditions respectively at other times, and accordingly the strata of the British Isles vary greatly when traced laterally. That the uplifts of the Fermo-Triassic periods produced some effect on the nature and distribution of the Jurassic rocks is certain, but it is not quite clear how far the ridges produced by these uplifts were submerged and denuded during the deposition of the main portion of the Jurassic strata.

Viewed broadly, the Jurassic rocks of Britain may be regarded as consisting of three great clay deposits, the

Lias, Oxford and Kimmeridge Clays, alternating with
the deposits of variable lithological characters, which
compose the Bajocian, Bathonian, Corallian, Portlandian
and Purbeckian subdivisions. This essentially argillaceous character of a large part of the deposits of
Jurassic age is often overlooked, as, owing to their sameness and the comparative paucity of organisms constituting the faunas in the clays, their description in text-
books can be given at much shorter length than that
of the more variable and highly fossiliferous deposits
which separate the clays. The following figure (Fig. 23)
roughly represents the nature of the different divisions
of the rocks of this system when traced across England
from south-west to north-east.

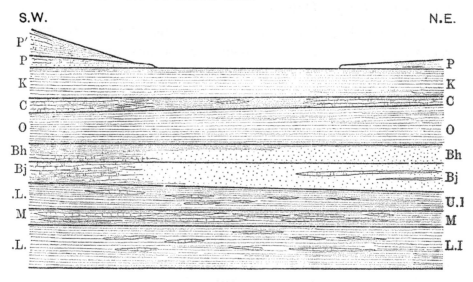

FIG. 23.

Vertical scale : 1 in. = about 1000 feet.

It will be seen that the greatest variations in lithological character occur in the Bathonian and Bajocian
beds, and it will be of interest to give some account of

the principal variations and to attempt to account for them. In so doing it will be convenient to consider the four major divisions of the Jurassic rocks separately, and to enter into particulars concerning the local classification applied to the rocks of these divisions.

The Lias. The British Lias deposits are divided into the Lower Lias, the Marlstone, and the Upper Lias corresponding in general terms only with the Sinemurian, Liassian, and Toarcian. The Marlstone is separated from the Upper and Lower Lias on account of the greater percentage of carbonate of lime which it contains, so that the bands of argillaceous limestone are much more marked in the Marlstone than in the upper and lower divisions, which consist chiefly of clay. The three divisions possess very much the same characters throughout the country, though the presence of the Mendip ridge and its continuation beneath London is marked by the attenuation of this and succeeding strata, and by the conglomeratic character of some of the Liassic strata where they abut against it. The British Lias, as a whole, seems to have been deposited in a fairly shallow sea at no great distance from the land. It passes down conformably into the Rhætic beds, indeed the zone of Ammonites (*Aegoceras*) *planorbis*, referred by British geologists to the Lower Lias is included by some continental writers with the Rhætic beds, and the plane of demarcation here as in other cases is conventional.

The Lower Oolites. Of all the British strata, these perhaps cause most trouble to the learner, on account of the different nomenclature applied to the rocks in different parts of England, and the rapid variations in lithological character, when the beds are traced laterally. The following divisions are usually adopted for the beds of the

south-western counties where the most marked marine
development occurs:—

> Cornbrash,
> Forest Marble,
> Great Oolite (with Bradford Clay),
> Fuller's Earth,
> Inferior Oolite.

Of these divisions, the uppermost one, the Cornbrash,
though thin, retains its characters with great constancy
across the island. Of the others the Forest Marble may
be looked upon as a local development of the upper
portion of the Great Oolite, and the Fuller's Earth is a
local deposit, so that the Inferior Oolite and Great Oolite
constitute the important divisions of the Lower Oolites.
The variations in the characters of the rocks may be best
shown in tabular form.

Gloucestershire, &c.	South Northamptonshire	N. Northampton-shire and Lincoln	Yorkshire
Cornbrash	Cornbrash	Cornbrash	Cornbrash
Great Oolite	Great Oolite (Upper part)	Great Oolite Clay Great Oolite Limestone Upper Estuarine	Upper Estuarine
....	Northamptonshire	....	....
		Series Lincolnshire Limestone	Series Scarbro' Limestone Middle Estuarine Series Millepore Oolite
Inferior Oolite	Sands	Lower Estuarine Series	Lower Estuarine Series
Upper Lias	Upper Lias	Upper Lias	Upper Lias

The dotted line shows roughly the division between Bathonian and
Bajocian.

The changes may be explained very simply if we leave out of account for the moment the development of Lincolnshire Limestone, with its equivalent the Scarbro' Limestone, and the Millepore series. The beds in Gloucestershire and other south-western counties are essentially marine; whilst in Northamptonshire and Lincolnshire estuarine conditions set in after the deposition of the Upper Lias, and continued throughout the deposition of the Bajocian and Lower Bathonian beds, being replaced by marine conditions during the formation of the Upper Bathonian strata, and still further north in Yorkshire the estuarine conditions generally prevailed throughout Bajocian and Bathonian times. These changes point to the existence of land towards the north. The general simplicity is modified by temporary prevalence of marine conditions twice over (during the deposition of the Millepore Oolite and the Scarbro' Limestone) in Yorkshire, and once (during the deposition of the Lincolnshire Limestone) in Lincolnshire.

Certain local deposits have not been noticed, but two of them merit brief reference. At the base of the Great Oolite of Oxfordshire is an estuarine deposit of finely laminated mechanical sediment mixed with calcareous matter known as the Stonesfield Slate, especially interesting on account of its fossils, while a bed with similar lithological characters but with a different fauna occurring at the base of the Lincolnshire Limestone (of Bajocian age) is termed the Collyweston Slate. Neither of these deposits is a slate in the true sense of the word, as they have not been affected by cleavage subsequently to their accumulation, but each has been somewhat extensively used for roofing purposes.

The Middle Oolites are much less complicated though

considerable variations arise with respect to the Corallian Rocks. The Oxfordian with Callovian consist chiefly of clay, though the Callovian of the south of England is represented by calcareous sandstone, with a peculiar fauna which seems to be represented in the lower part of the Oxford Clay further north, though this Callovian fauna has not been everywhere recognised.

The Corallian of the southern counties consists of limestones with calcareous grits, the limestones being often largely composed of the remains of reef-building corals, and a similar development of the rocks of this series is found in Yorkshire, while a local development of the same character is found at Upware in Cambridgeshire, though in the other parts of the Fenland counties the Corallian is represented by an argillaceous deposit with Corallian fossils known as the Ampthill Clay.

The Upper Oolites have a tolerably constant base, the Kimmeridge Clay, usually consisting of laminated bituminous argillaceous material, but the Portlandian and Purbeckian divisions vary greatly, and are only locally developed, though their absence in some parts of central England is no doubt due to unconformity.

The Portlandian rocks of the south of England consist of limestones and sandstones which pass further northward into shallower water mechanical deposits often charged with iron hydrate, and the beds disappear in Oxfordshire. The Purbeckian rocks of the south are also limited as regards area of exposure: they consist of estuarine deposits with some terrestrial accumulations of the nature of old surface soils. Representations of the Portlandian and Purbeckian beds are found in Lincolnshire and Yorkshire, as arenaceous deposits in the former county and argillaceous ones in the latter. Both are marine deposits of a northern type, developed elsewhere in northern European and circumpolar regions, and in these counties we find a complete passage from the Jurassic rocks through the Cretaceous rocks, but the exact lines of demarcation between the different series of the passage beds are difficult to define.

The foreign Jurassic rocks of Europe and of some parts of Asia strongly resemble in general characters those which have been described above as occurring in Britain. One of the most remarkable features of the Jurassic rocks as a whole, is the absence of the Lias over wide areas, the continental period which in Britain existed in Fermo-Triassic times is elsewhere frequently replaced by one of Liassic age.

The Jurassic and Cretaceous rocks are of interest on account of the evidence which they supply as to the existence of climatic zones in these periods, which run fairly parallel with those at present existing. The late Dr Neumayr in a paper already cited divides the world during later Mesozoic times into four distinct climatic zones, equatorial, north and south temperate and boreal zones (the corresponding austral zone is not known owing no doubt to the extensive sea of South Polar regions and our general ignorance of its lands). In Europe the Mediterranean Province belongs to the equatorial zone, the Middle European to the North temperate zone, and the Russian or Boreal to the Boreal zone. The last-named is marked partly by negative characters, the absence of certain Ammonite-genera and of coral reefs being noticeable, whilst the lamellibranch *Aucella* is very frequent. In the North temperate zone, certain Ammonite genera as *Aspidoceras* and *Oppelia* are abundant and there are also extensive coral-reefs. The Equatorial zone is marked by the Ammonite-genera *Phylloceras* and *Lytoceras* and by the *Diphya* group of *Terebratulæ*. It is of special interest to note that the fauna of the South temperate bears closer relationship to that of the North temperate than to that of the intermediate Equatorial zone.

Jurassic floras and faunas. The Jurassic flora is very similar in its characters to that of the Lower Cretaceous rocks, and the two taken together afford a decided contrast with that of later Palæozoic times, and also with that which succeeds them in the Upper Cretaceous rocks, which bears a marked resemblance to the existing flora. Cycads predominate, accompanied by conifers, and a fair number of ferns and Equisetaceæ.

The Jurassic fauna is specially noteworthy on account of the character of the vertebrata, but some notice of the invertebrates must also be taken. The abundance of corals in the Temperate zones has already been pointed out, but the mollusca form the bulk of the invertebrate fauna, lamellibranchs, gastropods and cephalopods being all abundant; of the last-named the ammonites and belemnites contribute most largely. The vertebrates include remains of fishes, amphibia, reptiles, birds and mammals. The Jurassic reptilia furnish representatives of some modern orders as the Chelonia and Crocodilia, but the most important orders are essentially characteristic of later Mesozoic times and their representatives abound in the Jurassic strata. These are the Sauropterygia (including the Plesiosaurs), the Ichthyopterygia (including the Ichthyosaurs), the Dinosauria, and the Pterosauria commonly known as Pterodactyls. No birds have hitherto been discovered in the British Jurassic rocks, but the Solenhofen Slate of Bavaria (of Kimmeridgian age) has furnished the celebrated *Archæopteryx macrura*, which is not only placed in a family but also in an order by itself, the order Saururæ. Many remains of mammals have been extracted from the estuarine deposits of Stonesfield, and the old surface soils of the Purbeckian beds; representatives of the Monotremata are furnished by the *Plagiaulacidæ* and *Tritylodontidæ*, the former family containing the genus *Plagiaulax* of the Purbeck Beds and the latter, *Stereognathus* of the Stonesfield slate. The Marsupialia are represented by the *Amphitheridæ, Spalacotheridæ* and *Triconodontidæ*. Some forms have been referred to the Insectivora, but there is still disagreement concerning the correctness of this reference.

Before dismissing the subject of the Jurassic fossils, attention may be called to a feature which has been frequently commented upon, namely, the general resemblance of the flora and fauna of Jurassic times to the modern Australian fauna and flora. The explanation which has been offered to account for this resemblance has been given in a preceding chapter, where it was stated that Mr A. R. Wallace considers, after review of the geological and biological evidence, that Australia was severed from the adjoining continental lands in Mesozoic times, and that the higher forms of life which on the larger continents have replaced the earlier and lower forms have not succeeded in obtaining a footing in Australia, which therefore furnishes us with a local survival of a once widespread fauna. In connection with this matter the actual existence of the genus *Trigonia* (a form peculiarly abundant in Jurassic strata and characteristic of Mesozoic strata in Britain) in the Australian sea is of considerable interest.

CHAPTER XXIII.

THE CRETACEOUS SYSTEM.

Classification. The rocks of the Cretaceous system are conveniently divided into Upper and Lower Cretaceous. The following classification has been widely used for the British deposits, and is founded on lithological characters:

Upper Cretaceous
- Upper Chalk with flints ⎫
- Middle Chalk with few flints ⎬ Chalk
- Lower Chalk without flints ⎪
- Chalk Marl ⎭
- Upper Greensand
- Gault

Lower Cretaceous
- Lower Greensand
- Wealden
- Hastings sands

As the result of examination of the faunas, a more generally applicable classification has been established and is now largely adopted. It is as follows:

- Danian ⎫
- Senonian ⎬ Upper Cretaceous
- Turonian ⎪
- Cenomanian ⎭

- Albian ⎫
- Aptian ⎬ Lower Cretaceous.
- Neocomian ⎭

In this classification the Neocomian practically represents the Wealden and Hastings beds, the Aptian the Lower Greensand and the Albian the Gault, placed according to this classification in the Lower Cretaceous, while the Upper divisions represent the strata above the Gault, consisting essentially of Chalk in England.

Description of the Strata.

(i) *The Neocomian and Aptian Beds.* In the south of England the Lower Cretaceous beds succeed the Jurassic rocks with little or no break, and the type of the lower beds is similar to that of the beds deposited during the Purbeck age, consisting of estuarine deposits of variable characters, chiefly arenaceous below (the Hastings sands) and argillaceous above (the Wealden series), though impure limestones are found, largely composed of the shells of the freshwater *Paludina*, and much ironstone is developed in places. At the close of Neocomian times, the freshwater conditions in southern England were replaced by marine conditions and the Lower Greensand strata with their marine fauna were deposited in the Aptian sea. The Neocomian and Aptian beds thin out westward, and much more rapidly to the northward, so that both divisions disappear against the now buried ridge which forms a continuation of the Mendip axis. North of this they appear in another form. At first the highest Aptian beds alone are developed as shore deposits. Passing into Norfolk lower beds come in until in Lincolnshire we get a complete development of the Neocomian and Aptian beds with a marine facies, though of fairly shallow water character, whilst in Yorkshire the two divisions are represented by a deeper water clay, forming the Upper portion of the Speeton series. There is a consensus of opinion in favour of the Neocomian beds of southern Britain having been laid down in an estuary of a river flowing from the west over a continent now destroyed. To the north of this river stood the London ridge of the Palæozoic rocks, the northern borders of which formed the coast line off which were deposited the sediments of Neocomian and Aptian ages which occur in northern England. Before the deposition of the Albian beds a considerable upheaval of some parts of Britain occurred, and an unconformity separates the higher Cretaceous beds from older strata of Cretaceous and Jurassic ages, thus complicating the major phases by local changes in the characters of the strata.

(ii) *The Albian and higher Cretaceous Beds.* The commencement of the deep-water phase of the third marine period may be said to occur in Albian times in Britain, reaching its maximum during the deposition of the chalk. The existence of a deeper sea towards the north of England is indicated by the characters of the Albian and newer strata. The Albian beds of gault consist of a stiff clay in southern England, replaced by coarser mechanical sediments towards the west. As one passes north from the London ridge (which exerted its influence in Albian times, after which it was finally buried in sediment) the gault thins out, and becomes gradually replaced by calcareous deposit when it is known as the red chalk which replaces the gault in northern Norfolk, Lincolnshire and Yorkshire.

A local unconformity separating the chalk and gault in parts of East Anglia points to another local uplift with its accompanying complications in the characters of the strata. After the uplift had ceased, general depression

must have occurred, and the various divisions of the
chalk were accumulated in a fairly open sea, though, for
reasons to be given presently, this was probably of no
great lateral extent, save when united with the open
ocean, probably in a manner similar to the connexion
between the Gulf of Mexico and the Atlantic.

The general variations in the lithological characters of
the various members of the Cretaceous system will
probably be rendered clearer by reference to the accompanying diagram (fig. 24) representing the variations when
traced across England from south to north[1].

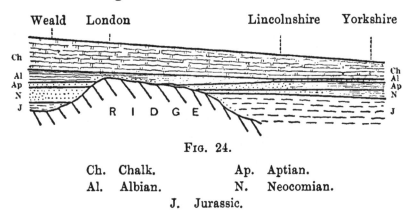

FIG. 24.

Ch. Chalk.   Ap. Aptian.
Al. Albian.   N. Neocomian.
J. Jurassic.

The clue to the physical geography of Britain during
Cretaceous times is furnished to a considerable extent by

---

[1] For information concerning the British Cretaceous beds, see Topley
and Foster, "Geology of the Weald," *Mem. Geol. Survey*, 1875; Bristow
and Strahan, "Geology of the Isle of Wight," *Mem. Geol. Survey*, 1889;
Lamplugh, "On the Speeton Clay," *Q. J. G. S.*, vol. xlv. p. 575, and
"The Speeton Series in Yorkshire and Lincolnshire," *ibid.*, vol. lii.
p. 179; Barrois "Recherches sur le Terrain Crétacé supérieur de
l'Angleterre et d'Irlande," Lille, 1876; and various papers by Messrs
Hill and Jukes-Browne, in the *Quarterly Journal of the Geological
Society* and *Geological Magazine* of recent years. For the Scotch deposits consult a paper by Prof. Judd, *Q. J. G. S.*, vol. xxxiv. p. 736,
and for those of Ireland, see Hume, *Q. J. G. S.*, vol. lii. p. 540.

study of the foreign deposits. In Northern Europe the Cretaceous beds of England are met with in Northern France, and there the characters are generally speaking. similar to those of our British deposits. In Germany shallower water conditions prevailed, the lower beds gradually disappear, and the upper beds are replaced by mechanical sediments of various degrees of coarseness, becoming on the whole coarser, as one travels eastward, so that in Saxony the chalk is partly replaced by arenaceous deposits (the 'Quader' sandstones) which are responsible for the remarkable scenery of the Elbe district above Dresden. In passing northwards, indications of similar change are noted in the deposits of Denmark and Scania, whilst to the south, we get a complete change in the character of the rocks, after crossing the Loire in France, and a similar change is observable in districts lying further east. Furthermore, as will be noted more fully in a subsequent paragraph, the character of the Upper Cretaceous flora indicates the existence of a large tract of land lying to the north and north-west of Europe, so that it would appear that the Cretaceous rocks of Northern Europe were deposited in a gulf-like expansion of a western ocean, bounded on the north by Scandinavia, on the west by eastern Germany, and on the south by a ridge running eastward from the mouth of the Loire[1]. We may speak of this gulf as the Chalk gulf. To the south of the presumed ridge the character of the strata alters, and also that of the included organisms. This southern type of Cretaceous rocks is one which is very widely spread, being found in Europe south of the Loire,

[1] The reader will find the existence of this gulf maintained and supported by a considerable mass of detail in Mr A. R. Wallace's *Island Life*.

and of the Alps, and in Greece and Turkey, while it also occurs in the northern parts of Africa. The beds of this type are traceable through Asia Minor into India and to the shores of the Indian Ocean, indicating the existence of a widespread Cretaceous ocean, which is sometimes spoken of as the Hippurite-limestone sea, for reasons which will eventually appear. The deposits are largely formed of hard limestone which is very different in its character from the soft chalk of the northern gulf.

The climatic conditions which prevailed during Cretaceous times were apparently similar in most respects to those of the preceding Jurassic period, and as already stated the climatic zones which Neumayr defined for Jurassic times are also maintained by him to have existed during the Cretaceous period. The existence of cold has sometimes been inferred from the presence of large foreign blocks in the chalk, especially at its base, but if these are due to the transport, they might well be caused by masses of floating ice, which are often found at considerable distances from the coast in temperate regions after the break-up of the frost which succeeds an unusually hard winter. The flora and fauna are not suggestive of severe conditions.

The Cretaceous flora and fauna. It has been noted in the last chapter that the gymnospermous flora of the Jurassic period, in which cycads form a considerable percentage of the whole flora, was prevalent in Lower Cretaceous times. In the Upper Cretaceous rocks this flora is replaced by one which consists to a large extent of dicotyledonous angiosperms. These are found in the Upper Cretaceous rocks of Europe and North America, and as the researches of botanists indicate their origin in circumpolar regions, their arrival in Europe is an additional argument in favour of the existence of an extensive northern continent, sending a prolongation to the southward in eastern Europe.

The invertebrate fauna bears considerable resemblance to that of Jurassic times, and many of the dominant Jurassic genera are also found in Cretaceous rocks. A most interesting feature is connected with the character and geographical distribution of the Ammonites. In Europe they are almost exclusively confined to the deposits of the northern gulf, and before their final disappearance they undergo many changes of form. We find the discoid spiral shells of earlier times, but these are accompanied by shells which are straight, curved, boat-shaped, and coiled into various helicoid spirals, sometimes having the whorls in contact, while at other times they are separate.

In the chalk of Britain gastropods are on the whole rare, and this fact serves to emphasize the palæontological break which occurs between the Cretaceous and Tertiary rocks; but when conditions were favourable, as during the deposition of some of the strata of the Middle Chalk, gastropods are abundant, and some are related to Tertiary genera, so that we may assume that the palæontological break alluded to is exaggerated by the difference of conditions which prevailed during the deposition of the earliest Tertiary and latest Cretaceous sediments.

In the Cretaceous deposits of the southern sea, where the Ammonite tribe is almost unknown, the remarkable family of the lamellibranchs known as the Hippuritidæ furnish the dominant invertebrates of the period, and the representatives of this family are exceedingly scarce amongst the Cretaceous strata of the northern gulf, though they are found on two or three horizons.

Of vertebrates, the most interesting are the reptiles.

The families which predominate in Jurassic times have many representatives amongst the Cretaceous strata also, but the order Squamata is represented by the sub-order Pythonomorpha, which is characteristic of the Cretaceous rocks. The best known representative is the gigantic *Mosasaurus*. Lastly, we have the remarkable toothed birds or Odontornithes, now placed in different orders, the genus *Hesperornis* being the only representative of the sub-order Odontolcæ of the Ratitæ, whilst *Ichthyornis* and allied forms are placed in the sub-order *O*dontormæ of the Carinatæ.

CHAPTER XXIV.

THE EOCENE ROCKS.

Classification. The Eocene Beds of the south of England have been subdivided according to the variations in their lithological characters, and the subdivisions have received local names. The following classification is generally adopted, though the different subdivisions are by no means of equal value:

Upper Eocene	Upper Bagshot Beds	
	Barton Beds	
Middle Eocene	Bracklesham Beds	
Lower Eocene	Lower Bagshot Beds	
	London Clay[1]	
	Oldhaven Beds	Lower London Tertiary Strata
	Woolwich and Reading Beds	
	Thanet Sands	

The deposits vary greatly when traced abroad, and the exact equivalents of the minor subdivisions of the British rocks can seldom be ascertained at any distance from England, though the division into Upper, Middle, and Lower Eocene can be made over wide areas.

Description of the strata. The character of the strata of Europe and Asia indicates the persistence of the

[1] Some writers place the London Clay in the Middle Eocene.

northern gulf and southern ocean of Cretaceous times in
Eocene times also, though the area of each had shrunk
in the meantime, owing to the physiographical changes
which occurred at the end of Cretaceous times, giving rise
to more extended land areas, and producing a shallow
water phase over wide extents of ocean,—the final
shallow water phase of the third and last great marine
period of the British area. It is difficult to ascertain the
exact importance of the physical break between Cretaceous and Eocene rocks in the south-east of England,
owing to the subterranean solution of the upper part of
the chalk, subsequently to the deposition of the Eocene
strata, but the contraction of the Cretaceous gulf is shown
in several ways, one of the most significant being the
distribution of Cretaceous and Eocene rocks in the south-
west of England. The existence of an outlier of Cretaceous rock at Buckland Brewer in North Devon, only
three miles from the Atlantic Ocean, indicates the former
extension westward of the Upper Cretaceous beds, while
the occurrence of an outlier of Eocene rocks at Bovey
Tracey in South Devon, resting not on Cretaceous but on
Palæozoic rocks, shows that there was an uplift after the
deposition of the Cretaceous rocks and before the Eocene
rocks were deposited there, and that during the period of
uplift the Cretaceous rocks were removed.

Owing to these physical changes, the Eocene rocks of
Britain are mainly mechanical sediments, some, as the
Oldhaven beds, being composed of coarse pebbles over a
fairly wide district, while some of the earlier Eocene
rocks are estuarine or fluvio-marine.

The Eocene rocks of Britain occur in four areas,
namely, the London Basin, the Hampshire Basin, the
Bovey Tracey outlier, and the north-east of Ireland and

western Isles of Scotland. The deposits of the thre
southern areas may be considered together, and giv
general indications of an approach to land when passin
westward. The Lower London Tertiary strata are fluvi
marine at the east end of the London Basin ; they becom
shallower water deposits when traced westward, and begi
to disappear. The London Clay is an estuarine deposi
which is generally supposed to have been laid down a
the mouth of a large river flewing from the west. It
absent in the Bovey Tracey outlier.

Local disturbances caused the existence of a shallo
water region in the east during the deposition of th
Middle and Upper Eocene deposits, and accordingly th
well-marked marine deposits which form the represent
tives of these divisions in Hampshire are replaced by th
Bagshot beds of the London Basin, consisting chiefly
coarse mechanical sediments with a poor marine faun
but even in the west shallow water prevailed at tim
during the accumulation of various plant-bearing strat
The Middle Eocene beds only are found in the Bove
Tracey outlier, though the Upper Eocene beds ma
originally have been laid down in that area, and subs
quently denuded.

The fourth area displays a very different successio
of Eocene strata, and one of extreme interest. Mechanic
sediments and plant-bearing clays and lignites alterna
with a vast accumulation of basaltic lavas, indicating th
outbreak of the volcanic forces in the British area, after
period of quiescence which lasted through the great
part of Mesozoic times. The region in which these lav
were poured out was probably a land area during th
greater part of the period of volcanic activity, but th
horizontal lie of the lava flows and their wide exte

indicate the existence of a flat tract of country, gradually raised into a plateau by the accumulation of sheet over sheet of basalt. How far this plateau extended it is impossible to say. The distribution of the lavas at the present day is somewhat limited in our isles, but there is no sign of dying out at the present margins of the accumulations, and they have probably escaped denudation in these regions, as maintained by Professor Judd, on account of the faults which have depressed them, while the portions which were not depressed have been removed by denudation. Two views as to the origin of the lavas have been put forward : according to Prof. Judd, they were poured forth from gigantic volcanoes, while Sir A. Geikie maintains that they represent portions of massive or fissure eruptions, the molten rock having welled out from great cracks in the earth, which are now filled by once molten rock in the form of dykes. As these dykes extend far away from the present volcanic plateau, one actually extending to the Yorkshire coast, we may well believe, whatever was the origin of the sheets of lava, that they were formerly spread far away from their present terminations[1]. Without entering here into a discussion of the exact nature of extrusion of these igneous sheets, it will suffice to say that all the evidence points to the formation of extensive plateaux, which must have presented a fairly uniform surface, similar to that which is still found characterising the volcanic districts of the western territories of North America.

[1] Prof. Judd's views will be found in a series of papers by him on the " Secondary Rocks of Scotland," *Quart. Journ. Geol. Soc.*, vol. **xxix.** p. 95, **xxx.** p. 220, **xxxiv.** p. 660, while Sir A. Geikie's explanation is advanced in a paper in the *Transactions of the Royal Society of Edinburgh*, vol. **xxxv.**; see also the same author's *Ancient Volcanoes of Great Britain.*

The Eocene rocks of the north-west of Europe possess characters very similar to those of the south of England, and there are indications that the northern gulf had diminished in extent towards the east as well as towards the west.

Passing to southern Europe, Central Asia and northern Africa, we find the conditions of Cretaceous times reproduced, and an extensive series of marine deposits extends very widely over these regions, the most persistent deposit being a mass of limestone of Middle Eocene age, which is almost entirely composed of the tests of Nummulites, whence the development is known as the Nummulitic Limestone facies, and we may speak of the ocean as the Nummulitic Limestone Sea. The incoming of shallow water conditions marked by accumulation of coarse mechanical sediments towards the end of the Eocene period in some parts of the southern European area indicates the setting in, even then, of those continental conditions which culminated during the Miocene period.

In North America we get similar evidence of the contractions of the oceans which in Mesozoic times occupied large expanses of our present continents.

The climatic conditions of Eocene times have been noticed in passing in chapter IX., and evidence was given to prove the prevalence of a warmer climate over the British area than that which now exists. A study of the floras of various parts of the northern hemisphere suggests that climatic zones, whose lines of demarcation ran practically parallel with the Equator, existed in Eocene times also, though further information upon this subject is desirable.

The Eocene flora and fauna. The flora of prevalent

dicotyledonous angiosperms, which appeared in Upper Cretaceous times, also marks the Eocene and later deposits, but a study of the floras indicates that the differentiation which now marks off the floras of different areas from one another had not occurred to so great an extent as at the present time. The existence of a rich flora in the Eocene beds of circumpolar regions in the northern hemisphere should be noted, though perhaps its importance has been somewhat exaggerated.

The invertebrate fauna shows an approximation to that of the present day. The remarkable ammonite fauna of Mesozoic times has disappeared, and gastropods and lamellibranchs predominate, many of the forms belonging to existing genera, though very rarely to existing species. The Nummulites are the most characteristic Eocene fossils, and the period may be spoken of as the Nummulitic Period, though it is now known that Nummulites are not confined to the Eocene strata.

The vertebrate fauna is very noteworthy. The fishes and reptiles are closely related to existing forms, and the orders of reptiles which predominated in Mesozoic times have completely disappeared. But the mammals are the most interesting vertebrates of the Eocene period. Instead of the lowly organised forms of Mesozoic times, we find representatives of many orders, including the highest, the Primates. The generalised forms which serve as links between groups which are now separated to a considerable extent are of particular importance. They have been detected in Eocene rocks of various regions, though the most complete series have been obtained from the Eocene rocks of North America and made known to us through the numerous memoirs of Professors Cope and Marsh[1].

[1] The Eocene floras of Britain are described by Mr J. Starkie

Gardiner and Baron von Ettingshausen in the *Monographs of the Palæontographical Society*; other Monographs of the same Society contain an account of the Eocene Mollusca by Mr F. E. Edwards and Mr S. V. Wood.   An idea of the generalised forms of Mammalia may be obtained by perusal of that portion of Nicholson and Lydekker's *Manual of Palæontology* in which the latter author treats of the Mammalia, and in this connexion the reader will do well to read Prof. Huxley's " Lecture on Fossil Horses," reprinted in his *American Addresses*.

CHAPTER XXV.

THE OLIGOCENE AND MIOCENE PERIODS.

(i) *The Oligocene Beds.*

Classification. The Oligocene Beds of Britain are classified as follows:—

Upper	Wanting
Middle	Hempstead Beds
Lower	Bembridge Beds
	Osborne Beds
	Headon Beds

Description of the strata. Little need be said of the deposits of this period, either in Britain or abroad, except to remark that they show the further spread of continental conditions over the regions now occupied by land. The British deposits are now seen in the Hampshire Basin only, and have been spoken of as the fluvio-marine series, as many of the strata were laid down in continental sheets of water, while the true marine sediments are thin and infrequent.

The lithological characters of deposits formed under these conditions naturally vary greatly, consisting of different kinds of mechanical sediments occasionally mixed with thin freshwater marls and limestones. On the Continent similar conditions prevailed, though the occurrence of fairly wide tracts of level surface is indicated by the widespread distribution of beds of brown coal or

lignite, and the coarse and thick Oligocene 'nagelfluh.' of
Switzerland points to the elevation of mountain ranges in
the neighbourhood.

The flora and fauna. The remarks made concerning
the Eocene flora and fauna are generally applicable to
those of Oligocene times, except that the Oligocene fossils
bear a still closer resemblance to living forms, and the
Nummulites are no longer dominant.

(ii) *The Miocene Period.* Beds of Miocene age are
wanting in Britain, and on the Continent they occur in
isolated basins deposited in gulf-like prolongations of the
ocean, never very far from land. A description of the
strata and their fossil contents would be of little use for
our present purposes, and the remarks made concerning
the Oligocene beds will apply to the Miocene strata
also.

The period was mainly remarkable on account of the
important physical changes which occurred, to which
we must devote some consideration. Commencing with
the British area, we find in the south evidence of the
separation of the London and Hampshire Basins at this
time, for the Oligocene beds of Hampshire are tilted up
on the south side of an anticline, which separates the
Hampshire Basin from that of London, indicating that
the movement was post-Miocene, while in Kent, beds
of Pliocene age rest on the denuded top of the chalk,
showing that the elevation and denudation which accompanied it were pre-Pliocene; the great Wealden
anticline is thus seen to be of Miocene age. On the
north side of the London Basin the line of demarcation
between Eocene and Mesozoic beds runs approximately
parallel to the strike of the latter in that part of Britain,
and this points to the elevation of the Mesozoic strata

which gave them their present south-easterly dip about the same period, though in the absence of Oligocene rocks it cannot be definitely stated that the movement was altogether post-Oligocene.   The present physical geography of considerable parts of Britain must date from Miocene times.

Important as the changes were in Britain, they were slight as compared with those which affected Europe and many parts of Asia.   The great mountain chains of the Old World received their maximum uplift during this great period of earth-movement, and orogenic structures were impressed upon the rocks of many regions, for the Tertiary Mountain Chains of the Old World have an Alpine structure impressed upon them as the result of intense lateral pressure, accordingly we find the Eocene strata lifted far above their original level to heights of 8,000 feet in the Alps and over 12,000 feet in the Himalayas.   Away from these marked uplifts epeirogenic movements caused the disappearance of the seas of earlier Eocene times, so that towards the close of the Miocene Period, the main features of the Eurasian continent were much as they are now.   The present drainage-systems must have originated at the same time, and the sculpture of our continent has been carried on more or less continuously by subaerial agents from Miocene times to the present day.   That any addition to the total area of land was made is doubtful.   The land which appears to have existed to the west of Britain during Cretaceous and Eocene times finally disappeared beneath the waters of the Atlantic Ocean, and the movement probably gave rise to the prominent submarine feature which now exists at some distance from the coast of Ireland.   A great marine period is now existent in our ocean areas, but so far as the existing continents are concerned, we are living on the fourth continental period which practically came into existence in Miocene times.

The strike of the uplifted strata naturally coincides on the whole with the axes of the major uplifts, and accordingly we find the Mesozoic and early Tertiary strata folded around axes which have a prevalent east and west direction, with others which have a trend at right angles to this. The strike of the British Mesozoic rocks seems to have been determined by each of these sets of movements, so that although it is east and west in the south of England, it runs north and south in the eastern counties north of the Thames.

In America, although epeirogenic movements had occurred before Miocene times, with the formation of wide continental tracts, these appear to have been of the nature of plains, diversified by extensive inland sheets of water, and uplift of orogenic character converted these plains into uneven tracts in Miocene times. Many of the movements in America, which like those of Europe are still progressing with enfeebled power, differ from those of Eurasia, giving rise to raised monoclinal blocks rather than to violent folds of Alpine character, as seen in the western territories of North America, and as proved also by the differential movements which are now known to affect the Atlantic coast of that continent.

Accompanying these changes in the earth's crust were others which affected the climate, at any rate locally. The warm climate of Eocene times gradually gave way to a cooler climate in Oligocene times, and this lowering of temperature was still further advanced in Miocene times, though there is evidence that the temperature of those parts of Europe which have strata representative of the Miocene period was higher than it is at the present day.

Owing to the changes which occurred in Miocene times, the area of sedimentation was extensively shifted to our present oceans, and accordingly we find that the times subsequent to those of the Miocene uplifts are marked by scattered accumulations of continental character, with a few insignificant marine strata seldom found far inland from the the present coast-lines.

CHAPTER XXVI.

THE PLIOCENE BEDS.

Classification. The Italian Pliocene Beds which have long been known have been divided into three stages, to which names have been applied which are somewhat widely used, though the division of the British deposits into the same three stages has not been made. The stages are :—

> Astian.
> Plaisancean.
> Zanclean.

The classification of the British deposits may be made as follows :—

> Cromer "Forest" Series.
> Weyborne Crag and Bure Valley Beds.
> Chillesford Crag.
> Norwich Crag and Red Crag.
> Upper Coralline Crag.
> Lower Coralline Crag.

As the English deposits are somewhat scattered it is difficult to make out the exact order of succession, but the above shows the classification which is adopted by the best authorities, the Norwich Crag (or Fluvio-marine Crag as it is sometimes termed) being now supposed to represent the upper portion of the Red Crag.

Description of the strata. The British deposits are chiefly found in the counties of Norfolk and Suffolk, but isolated patches have been detected in Kent and at St Erth in Cornwall; while the inclusion of Pliocene fossils in the glacial deposits of Aberdeenshire and on the west coasts and islands of Great Britain suggests the occurrence of Pliocene beds beneath sea-level, around the British coasts, at no great distance from the land.

The term 'Crag' has been applied to shelly sands of which the British Pliocene beds are largely composed. The oldest British Pliocene strata are supposed to be the Lenham Beds, occurring in 'pipes' on the Chalk of the North Downs, which are referred to the Lower Coralline Crag, and some writers believe that the St Erth beds of Cornwall are of similar age[1]. The former are ferruginous sands, and the latter shelly sands and clays. The higher beds of the Coralline Crag are found in Suffolk, and are largely calcareous, being made of remains of polyzoa, molluscs, and other invertebrates. They were probably deposited in deeper water than the rest of the British Pliocene strata, and contain a far larger percentage of carbonate of lime. The Red Crag consists of ferruginous shelly sands, of the nature of sand-banks, formed near land; while the Norwich Crag is of a still more littoral character, and contains remains of land shells and the bones of mammalia mingled with the marine shells of the coast. The higher Pliocene deposits are also coastal accumulations, even the so-called Forest bed being a deposit and not a true surface soil, as proved by the observations of Mr Clement Reid. At the summit of the Cromer 'Forest' Series, however, is a true freshwater bed.

---

[1] See Clement Reid, *Nature*, 1886, p. 342; and Kendall and Bell, *Quart. Journ. Geol. Soc.*, vol. XLII. p. 201.

These British deposits appear to have been laid down
a coast line which formed one side of the estuary o
large river, of which the present Rhine is the 'betrunke
portion (to use a term introduced by Prof. W. M. Davis

On the European continent, marine Pliocene beds :
found in Belgium and Italy. The former deposits grea
resemble our Crags, whilst the latter are of interest
account of the mixture of volcanic beds with mar
sediments in Sicily, showing that the formation of E1
commenced in Pliocene times. Various deposits forn
in inland basins are found in France and Germany, 1
the most remarkable occur in the Vienna basin, wh
Caspian conditions prevailed over large areas, and 1
ordinary strata alternate with chemical deposits of wh
the best-known are the celebrated rock salt masses
Wieliczka, near Cracow. At the same time volca
activity was rife to the south of the Carpathian mountai
Other deposits, which are partly referable to the Plioce
period, occur in Greece at Pikermi, and in India in ·
Siwalik hills; these are celebrated for their remarka
mammals, as are the Pliocene strata of the West
territories of North America. The occurrence of marl
earth-movements since Pliocene times is indicated by
nature of the deposits of Barbadoes, where radiolar
cherts have furnished two echinids which are described
Dr Gregory as deep-sea forms. These beds were o
referred to the Miocene period, but there is good rea
for assigning them to a later date, and correlating tl
with the Pliocene beds of other areas, in which case tl
must have been a considerable uplift in this re{

---

[1] See a paper by Mr F. W. Harmer, "On the Pliocene Deposi
Holland, and their relationship to the English and Belgian Cr
Quart. Journ. Geol. Soc., vol. LII. p. 748.

since Pliocene times, a fact of great theoretical importance.

The climatic conditions of Pliocene times show steady fall of temperature. The early Pliocene beds of Britain were deposited during the prevalence of warmer temperatures than those which now exist in the same area, but during later Pliocene times, the temperature was at first similar to that now prevailing, and afterwards distinctly colder, and we find in the upper Pliocene beds the remains of organisms of a northern type. In the uppermost deposit of the Cromer 'Forest' Series, the arctic birch and arctic willow indicate the commencement of the cold which culminated in the succeeding 'Great Ice Age.'

The flora and fauna. Little need be said of the Pliocene fossils: the flora approaches that of present times, and the invertebrates are in most cases specifically identical with those now living. The vertebrates alone differ markedly from living forms, being chiefly of extinct species, and in many cases belonging to extinct genera. It is interesting to find that the mammalian fauna of Pliocene times resembles the existing fauna of the area in which the beds are found, a fact long ago observed by Darwin. Thus the European Pliocene mammals are like existing European forms, whilst in Australia the mammalian terrestrial fauna consists of Marsupials, and in South America there are Edentata of Pliocene age[1].

[1] The Pliocene fauna of Britain is described by Mr Searles V. Wood in the *Monographs of the Palæontographical Society.*

CHAPTER XXVII.

THE PLEISTOCENE ACCUMULATIONS.

Classification. The term Pleistocene, as used here, is approximately equivalent to the expressions 'Glacial Period' and 'Great Ice Age' of some writers; but I have adopted it in preference to these expressions, because it may eventually be possible to define the Pleistocene period in such a manner as to give the term a strictly chronological meaning, whereas the other terms indicate the existence of climatic conditions which must have ceased in some areas sooner than in others. At present, climatic change gives us the best means for separating the accumulations formed subsequently to the Pliocene period over large parts of the Eurasian land-tract, and the most convenient division of these continental accumulations is to refer them to three periods, viz. :—

The Forest Period (in which we are now living).
The Steppe Period.
The Glacial Period.

Some of the accumulations which were formed during the Steppe period are included in the Pleistocene period by many writers, but I prefer to treat of them as Post-pleistocene.

In the present state of our knowledge of the glacial deposits any attempt to make a classification applicable over very wide areas is doomed to failure, and the very principles upon which the classification should be based are a subject of disagreement. The most promising basis for classification is founded on alternate recession and advance of land-ice, though the proofs that advance takes place simultaneously over very wide areas are not yet forthcoming. Dr J. Geikie in the last edition of his work *The Great Ice Age* adopts four periods of glaciation, with intervening periods of recession, and this division accords with the observations of many foreign geologists. In order to understand the method of classification upon this basis, a few words concerning glacial deposits in general will not be out of place. Glacial accumulations may be divided into three classes:—(i) true glacial accumulations, formed on, in, and under the ice, and left behind upon its recession, (ii) marine glacial deposits, laid down in the sea, when floating ice is extensively found on its surface, and (iii) fluvio-glacial deposits, laid down by streams which come from the ice. The two former indicate glacial conditions, while the occurrence of fluvio-glacial deposits overlain by true glacial deposits indicates an advance of land-ice, for the fluvio-glacial deposits are accumulated in front of those which are truly glacial. Accordingly if we find alternations of glacial and fluvio-glacial deposits on a large scale, we may fairly infer the alternation of periods of great glaciation with others when the ice diminished, or in other words of glacial and interglacial periods. There is, however, in many cases great difficulty in distinguishing glacial deposits from marine glacial ones, while some of the true glacial deposits formed *in* the ice (englacial deposits) cannot readily be distinguished from those of fluvio-glacial origin. Furthermore, as the terminal moraines of land-ice

often rest upon other true glacial deposits, it is of
difficult to know whether we are dealing with the produ
of one or two glaciations over limited areas. The t
of superposition is often applicable, and one is enabled
obtain some clue as to the relative order of events.
England at least three periods of glaciation seem
be indicated by the glacial deposits. On the east co
the Cromer Forest Series is succeeded by the Cron
Till, and in Yorkshire the Basement Clay occupies
similar position with regard to the overlying glac
accumulations to that of the Cromer Till. Whetl
these deposits be marine or terrestrial, and the evidei
is not yet sufficient to settle this question to the sa·
faction of all geologists, there is no doubt that they :
glacial. Above them, in East Anglia, lies the Contor·
Drift, the origin of which is still a moot point, and it
overlain by the great Chalky Boulder Clay, which exter
far and wide over East Anglia, the Midland Counties a
into Yorkshire. Evidence has been adduced to conn
this with the *till* or boulder clay which spreads over ·
upland districts of the north of England at the foot of ·
main hill-systems. This set of deposits indicates a sec·
glaciation. As the upland till is often ploughed out
glaciers which have left their traces in the form
moraines in our upland regions, we seem here to h
evidence of a third glaciation, which naturally lea
no traces in the southern districts, and the exact age
this cannot be ascertained in the absence of fossil evidei
though we may provisionally refer it to the Pleistoc
period.

Another attempt has been made to classify the gla
deposits, on the supposition that there have been per
of elevation and depression of the land during Pleistoc

times. Some writers advocate one interglacial period when the land was depressed to an extent of 1400 and perhaps 2000 feet, while others have advocated the occurrence of a number of such inter-glacial marine periods. The evidence for the supposed oscillations is furnished by the existence of shell-bearing sands associated with boulder clays at high levels, the best known being on Moel Tryfan in Caernarvonshire, near Macclesfield in Cheshire, and near Oswestry in Shropshire. As many geologists believe that these shells have been carried to their present position by ice in a way which it is not our province to discuss here, we may dismiss this method of classification as based upon events which cannot be proved to have occurred. In the present state of our knowledge, it is indeed best to avoid, as far as possible, classifications which are intended to be applicable over wide regions, and to devote our attention to local details, gradually piecing together the evidence which is obtained as the result of exhaustive examination of each separate area[1].

[1] The glacial literature of our own island only, is so extensive that the student may well be bewildered when he attempts to grapple with it. He is recommended to read the following general works :

J. Geikie, *The Great Ice Age.* 3rd Edition, 1894.

H. Carvill Lewis, *The Glacial Geology of Great Britain and Ireland.* 1894.

G. F. Wright, *Man and the Glacial Period*, 1892, and *The Ice Age in North America*, 1890.

Sir C. Lyell, *Antiquity of Man.* 4th Edition, 1873.

For the glacial geology of special regions the following papers may be consulted :

The Lake District and adjoining neighbourhood. R. H. Tiddeman, "Evidence for the Ice Sheet in North Lancashire &c." *Quart. Journ. Geol. Soc.*, vol. xxviii. p. 471. J. G. Goodchild, "Glacial Phenomena of the Eden Valley &c." *Quart. Journ. Geol. Soc.*, vol. xxxi. p. 55, and J. C. Ward, *Mem. Geol. Survey*, "The Geology of the Northern half of the Lake District."

The foregoing remarks will convince the student that any attempt to show the distribution of land and sea during any part of the glacial period is not likely to meet with general acceptance, as so much depends upon the terrestrial or marine origin of the deposits of the lowlands, and the mode of formation of the shell-bearing drifts of high levels. The occurrence of elevation to a greater height than that which our country at present possesses during portions at any rate of the glacial period has been inferred on general grounds, but direct evidence in favour of it is furnished by the existence of a number of ancient valleys on the land around our coasts, whose floors are often considerably below sea-level, while the valleys are now completely filled up with glacial accumulations, except where they have been partially re-excavated by streams which for some distance run above the courses of the ancient streams.

The climatic conditions of glacial times can only be briefly touched upon in this place. If the periods of

Yorkshire. G. W. Lamplugh, " Drift of Flamborough Head," *Quart. Journ. Geol. Soc.*, vol. XLVII. p. 384.

Lincolnshire. A. J. Jukes-Browne, *Quart. Journ. Geol. Soc.*, vol. XXXV. p. 397 and XLI. p. 114.

East Anglia. Clement Reid, *Mem. Geol. Survey*, " The Geology of the district around Cromer."

North Wales. T. McK. Hughes, " Drifts of the Vale of Clwyd " &c. *Quart. Journ. Geol. Soc.*, vol. XLIII. p. 73, and A. Strahan, " Glaciation of South Lancashire, Cheshire, and the Welsh Border, *ibid.*, vol. XLII. p. 486.

Switzerland. C. S. du Riche Preller, " On Fluvioglacial and Inter-glacial Deposits in Switzerland," *Quart. Journ. Geol. Soc.*, vol. LI. p. 369 and " On Glacial Deposits, Preglacial Valleys and Interglacial Lake formations in Sub-Alpine Switzerland," *ibid.*, vol. LII. p. 556.

The reader will find references to other works on the Glacial Geology of other districts by consulting the general works referred to on the pre-ceding page.

advance can be proved to be contemporaneous over wide areas, this points to alternations of colder and warmer periods, or at any rate of drier and wetter periods, though local advance may be due to a number of causes. It must be borne in mind that with the temperature remaining the same, advance of ice can be brought about by increased precipitation of aqueous vapour in the form of snow.

The question of the cause of the glacial period is one that only indirectly affects the stratigraphical geologist until he has accumulated sufficient evidence to indicate the cause. It must suffice to observe that the extremely plausible hypothesis of Croll (for which the student should consult Dr Croll's *Climate and Time*) does not explain the apparent gradual lowering of climate throughout Tertiary times till the cold culminated in the Pleistocene period, and the student will do well to remain in suspense concerning the cause of the Ice Age until further evidence has been brought to bear upon it.

The glacial flora and fauna. The glacial deposits naturally yield few traces of life, except those which have been derived from other deposits, and we are dependent for our information concerning the fauna and flora of the glacial period upon the remains furnished by the inter-glacial deposits. Unfortunately it is very hard to ascertain which deposits are interglacial, and many which have been claimed as such are either preglacial or post-glacial. The meagre evidence which we possess points to the existence of an arctic fauna or flora in Britain during the prevalence of this glacial period. A question which has received much attention of recent years is that of the existence of preglacial or interglacial man, on which much has been written. The existence of man in glacial times is probable, but it is the opinion of many of those who are most competent to form a judgment, that it has not been proved in the only conclusive way, namely, by the discovery of relics of man in deposits which are directly overlain by glacial deposits, or which at any rate äre demonstrably older than glacial deposits[1].

[1] On the question of preglacial and interglacial man, see W. Boyd Dawkins, *Early Man in Britain*; H. Hicks, *Quart. Journ. Geol. Soc.*, vol. XLII. p. 3, XLIV. p. 561, and XLVIII. p. 453; T. McK. Hughes, *ibid.*, vol. XLIII. p. 73; Sir J. Evans, *Presidential Address to British Assoc.* 1897.

CHAPTER XXVIII.

THE STEPPE PERIOD.

THE occurrence of a period marked by dry climate over wide areas of the Eurasian continent, and possibly also in North America, is evidenced by the widespread distribution of an accumulation known as *loess*, concerning the origin of which there has been much difference of opinion, though that it was formed subsequently to the glacial period seems to be generally admitted, inasmuch as it is largely composed of rearranged glacial mud. The formation of the loess as a steppe-deposit was first advocated by Baron von Richthofen, and his views were supported by Nehring after study of the loess-fauna. Richthofen's explanation of the loess as due to the spread of dust by wind in a dry region is becoming widely accepted, and it necessitates the widespread occurrence of steppe conditions, as the loess has a very extensive geographical range, and may be truly regarded as the normal continental deposit of Eurasia during the period immediately succeeding the glacial period. In our own country, as the sea cannot have been far distant during these times the normal loess is not found, but several accumulations occur, which on stratigraphical and palæontological grounds must be regarded as synchronous with the formation of the loess. These are certain rubble-drifts of the southern counties, the older river-gravels of southern England, and some of the older cave deposits of various parts of England. It is doubtful whether any classification into minute subdivisions can be adopted for them, though Prof. Boyd Dawkins has advocated their separation into an older age of River Drift Man, and a newer period of Cave Man, on account of the evidences of a lower state of civilisation afforded by examination of the River Drift implements when compared with those fashioned by Cave Man. Roughly speaking, the Steppe period corresponds with the period during which Palæolithic man existed, at any rate in north-west Europe, and we may speak of the Steppe period as the Palæolithic period, without asserting that Palæolithic man necessarily disappeared at the time when the climate changed and caused the replacement of Steppe conditions by others favourable to forest-growth.

Description of the accumulations. The loess consists of unstratified calcareous mud or dust, with a peculiar vertical fracture, and is interesting rather on account of the nature of its fossils and of its distribution than for its lithological characters. As it is not found in Britain it is not necessary to say much about it, but merely to refer to the published descriptions[1].

The British deposits require some notice, as their characters and mode of occurrence are of some significance. Along the south coast are deposits of coarse rubble which have yielded some organic remains, which have been described by Mr Clement Reid[2], who also discusses their

---

[1] An account of Richthofen's views by that author will be found in the *Geological Magazine*, Dec. 2, vol. IX. (1882), p. 293, and the fauna of the loess is described by Nehring (*Ibid.*, p. 570).

[2] C. Reid, "Origin of Dry Chalk Valleys and of Coombe Rock," *Quart. Journ. Geol. Soc.*, vol. XLIII. p. 364.

origin. The rock, also known as the Elephant Bed, consists of angular fragments of flint and chalk, and seems to have been produced by streams which were able to flow over the surface of the chalk when it was frozen. Many other similar deposits in the south of England, which are found on the open surface, may have had a similar origin.

The Palæolithic river-gravels are found at various distances above present river-levels, and are the surviving relics of alluvial deposits which were laid down when the rivers ran at a higher level than they now do. That they are newer than the main glacial drifts of the region in which they occur is indicated by the frequent presence in them of boulders derived from the drift. Their antiquity is shown by the physical changes which have occurred since their deposition (there having been sufficient time since then to allow of the excavation of some river-valleys to a depth of over one hundred feet beneath their former level), and also by the character of the included mammals which will presently be referred to. The deposits vary in coarseness, like those of modern alluvial flats, from the coarse gravels of the river-beds to the fine loams and marls of the flood-plains. They are found, in Britain, with their typical mammalian remains, south-east of a line drawn from the mouth of the Tees to the Bristol Channel.

The cave-deposits have a wider distribution than those which have just been noticed, being also found to the north-west of the above-mentioned line in Yorkshire, and in North and South Wales. In the south of England they are found as far east as Ightham in Kent, and in a westerly direction to Torquay and Tenby. The Ightham deposits occur in fissures and consist of materials which were apparently introduced from above by river action[1]. The cave-deposits of limestone areas are sometimes found in fissures, but at other times in caverns with a fairly horizontal floor, on which the various accumulations lie in order of formation. The deposits vary in character and may be divided into three groups, though accumulations of intermediate character are found; the first group consists of cave-earths and cave-breccias—formed by weathering of the limestone, and the retention of the insoluble residue, as a more or less ferruginous mud, mixed with angular fragments of limestone, and with the remains of creatures which inhabited the caves; the second group consists of true deposits laid down under water, as gravels, sands, and laminated clays; while the third is composed of limestone deposited from solution in water, in the form of stalagmite[2].

The organic contents of the Palæolithic period are of much interest, and it is desirable to discuss their character before making further observations upon the physical conditions of the period.

The Palæolithic flora and fauna. The plants of some of the earlier deposits of the age we are considering show the prevalence of cold conditions during their accumulation, for instance the Arctic birch and Arctic willow are

---

[1] The Ightham fissures and their contents are described by Messrs Abbot and Newton, *Quart. Journ. Geol. Soc.*, vol. L. pp. 171 and 188.

[2] The reader should consult Prof. W. Boyd Dawkins' works on *Cave Hunting* and *Early Man in Britain*, for information concerning the Cave Deposits. See also Sir C. Lyell, *Antiquity of Man;* Sir J. Evans, *Ancient Stone Implements of Great Britain*, and Sir J. Lubbock, *Prehistoric Times.* In these works references will be found to papers by Messrs Pengelly, Magens Mello, Tiddeman and others on the Caves of Devon, Derbyshire and Yorkshire. References have already been made to papers upon the Caverns of North Wales.

found in the accumulations beneath the implement-bearing Palæolithic deposits of Hoxne in Suffolk[1]. The invertebrate fauna consists essentially of the remains of molluscs. The loess molluscs are chiefly pulmoniferous gastropods which lived upon the land, though swamp forms are occasionally associated with them. The palæolithic river-gravels have yielded numerous land- and freshwater-molluscs of living species, though some which are abundant in the British gravels are now extinct in Britain, e.g. *Cyrena* (*Cobicula*) *fluminalis* and *Unio littoralis*. Marine deposits of this age are occasionally found, as at March, in Cambridgeshire, where the fauna closely resembles that of our present sea-shores.

The vertebrate remains are much more remarkable, and it is not quite clear that the association of forms whose living allies now live under widely different conditions has been satisfactorily explained. The river-gravels and cave-deposits contain remains of temperate forms, as the bison, and brown bear, associated with those of northern forms, as the mammoth, woolly rhinoceros, glutton, reindeer, and musk ox, and also with those whose living allies are inhabitants of warmer regions, like the lion, hyæna, and hippopotamus. One of the most remarkable creatures is the sabre-toothed lion or *Machairodus*, remains of which have been discovered in Kent's Cavern, Torquay, and in the caves of Cresswell Crags, Derbyshire.

The loess fauna consists of characteristic steppe animals, such as the jerboa, Saiga antelope and steppe-porcupine, and it is interesting to find an indication of this fauna in the Ightham fissures.

---

[1] These beds are described by Messrs Reid and Ridley, *Geol. Mag.* Dec. III. vol. v. p. 441. See also C. Reid on the "History of the Recent Flora of Britain," *Annals of Botany*, vol. II. No. 8, Aug. 1888.

The first undoubted relics of mankind are found i
the Palæolithic deposits, which are very widely sprea
over the Eurasian continent. They consist mainly (
implements of bone and stone, the latter being chippe(
but never ground or polished, though both bone an
stone implements are frequently ornamented with er
graved figures. The cave-deposits have furnished imple
ments of a higher type than those usually found in th
river-drifts, but the latter are also found in caverns i
deposits beneath those containing the higher type, henc
the division of the period into two minor periods, that (
river-drift man, and that of cave-man[1].

There are several questions of interest connected wit
the Palæolithic fauna, three of which deserve some notic
here. The absence of the relics of the Palæolithic marr
malia and of the human implements in the river-gravel
north-west of the line drawn between the Tees and Bristo
Channel, and the presence of the mammalian remains i
the caverns of that area requires some explanation. On
such explanation assumes that the relics were destroye
in the open country to the north-west of that line, owin
to glaciation, but it is not by any means universall
accepted.

Another difficulty which in the opinion of some write:
has not been fully cleared up is the mixture of apparentl
southern forms like the Hippopotamus, with others
northern character like the Musk ox, under such cond
tions as to show that the creatures lived in the Britis

---

[1] Concerning this matter, the reader should consult Prof. Bo;
Dawkins' *Early Man in Britain.* Sir J. Prestwich has argued in favo
of the existence of a group of implements found on the plateau south
the Thames of an age antecedent to that of the ordinary river-dr
implements. See *Quart. Journ. Geol. Soc.,* vol. xlv. p. 270.

area contemporaneously. Seasonal migration might account for it, but the wide belt of overlap of apparent northern and southern forms requires something more, though secular changes of climate might shift the belt of seasonal overlap from one place to another, causing the entire belt of overlap to extend over a considerable distance.

The third, and perhaps most important difficulty is the abrupt change from the Palæolithic type of implement to the Neolithic type, characteristic of the next period. Some implements, as those of the kitchen-middens of Denmark, and those found at Brandon and Cissbury in this country, have been appealed to as intermediate in character, but evidence has been brought forward to show that each set is truly Neolithic, the one being the implements of the lowly fisher-folk who lived contemporaneously with the makers of the highly finished polished implements of Denmark, while the others are unfinished implements thrown away during the manufacture on account of flaws or accidental fractures. The difficulty is increased when we take into account the great physical and faunistic changes which occurred between Palæolithic and Neolithic times.

The country was undoubtedly more elevated than it is at present during portions if not during the whole of Palæolithic times, as shown by the appearance of the great mammals in Britain, the discovery of their remains beneath sea-level, and especially the occurrence of remains in the caverns of rocky islands such as those of the Bristol Channel, where they could not possibly have existed unless the present islands were connected with the mainland.

The fossils of the times between the Glacial period and

the Neolithic period indicate variations of climatic cond
tions.   Upon this point I cannot do better than quote th
words of Sir John Evans in his Presidential Address t
the British Association at Toronto[1]. "At Hoxne th
interval between the deposit of the Boulder clay and (
the implement-bearing beds is distinctly proved to hav
witnessed at least two noteworthy changes in climat(
The beds immediately reposing on the clay are charac
terised by the presence of alder in abundance, of haze
and yew, as well as by that of numerous flowering plant
indicative of a temperate climate very different from tha
under which the Boulder clay itself was formed. Abov
these beds characterised by temperate plants, comes
thick and more recent series of strata, in which leaves (
the dwarf Arctic willow and birch abound, and which wer
in all probability deposited under conditions like those (
the cold regions of Siberia and North America.

    "At a higher level, and of more recent date than thes
—from which they are entirely distinct—are the bed
containing the Palæolithic implements, formed in a
probability under conditions not essentially different froi
those of the present day."

[1] *Report Brit. Assoc.* for 1897, p. 13.

CHAPTER XXIX.

THE FOREST PERIOD.

SUBSEQUENTLY to Palæolithic times, the physical conditions over Eurasia changed greatly, and at the commencement of Neolithic times the conditions were favourable for the growth of forests over wide regions of that continent. At the commencement of the Forest period the physical conditions were very much the same as they are at present, though minor changes have of course taken place since then, including probably a submergence of large parts of Britain to a depth of about fifty feet beneath its former level, as indicated by the existence of Neolithic submerged forests round many parts of our coast-lines.

The Forest period may be best subdivided for local purposes by reference to the civilisation of mankind at different times, and in this way we obtain the following divisions:

> Historic Iron age.
> Prehistoric Iron age.
> Bronze age.
> Neolithic age.

A classification may also be based upon changes in the flora. In Denmark the peat deposits of this age are

divisible into five layers, characterised by different dominant forms of trees. These are as follows in descending
order:

Fifth layer: Beech..............Iron age
Fourth layer: Alder
Third layer: Oak................Bronze age
Second layer: Scotch Firs......Neolithic age
Lowest layer: Poplar.

In our own country the forest growth has been much
interfered with by man, but the lower fenland peat gives a
good example of the material formed by forest growth. It
is not necessary to touch on the various accumulations
which are now being formed in different parts of our
island, except to remark that the deposits of the Forest
period give indications of earth-movements on a small
scale, which is well seen in the fenland, where the forest
peat is covered in places by a "buttery clay" with *Scro-
bicularia piperata* indicating submergence, and above this
is a marsh peat.

The flora and fauna of the Forest period are practically
those of the present day, though the larger forms of
mammalia have disappeared one by one. The Irish elk
and *Bos primogenius* probably became extinct early in the
period, while as far as Britain is concerned the wolf, bear,
and beaver have disappeared within historic times.

The relics of man deserve passing notice. The Neolithic period is characterised by the absence of metal
instruments, though those made of stone were much more
highly finished than those of Palæolithic times, and were
often ground and polished. The first metal which was
largely worked was bronze, which gradually replaced
stone, though stone was extensively used in the Bronze
age, as indicated by the imitation of bronze implements

in stone. The Bronze age in turn was replaced by the Prehistoric iron age; at first, when iron was scarce, bronze implements were merely tipped with iron, but ultimately the one metal was practically replaced by the other.

The date of the Palæolithic period is unknown; no approximate date can be satisfactorily assigned to it, but various calculations, founded on different data, have been made as to the age of the Neolithic period, and several of them agree in placing it at about 7000 years from the present time.

It will be seen that no sudden and violent change marks the incoming of the human race, which to the geologist is but one of a large number of events which have followed each other in unbroken sequence, and accordingly the thread of the story where abandoned by the geologist is taken up by the antiquary, and passed on by him to the historian[1].

[1] The student may obtain information concerning the Neolithic age in Britain in Boyd Dawkins's *Early Man in Britain;* Sir J. Evans' *Early Stone Implements of Great Britain*, and Sir J. Lubbock's *Prehistoric Times*. In the latter work he will find a good account of the Neolithic remains of Denmark and of the Swiss Lake dwellings. For information concerning the Bronze age he should consult Evans' *Ancient Bronze Implements of Great Britain*. The varied Danish antiquities of Neolithic and Bronze ages are figured in H. P. Madsen's *Antiquités Préhistoriques du Danemark*. The Prehistoric fauna of the fenlands is described in Sir R. Owen's *History of British Fossil Mammals and Birds*.

CHAPTER XXX.

REMARKS ON VARIOUS QUESTIONS.

THERE are many problems connected with geology which can only be solved by detailed study of the stratified rocks, and when solved the principles of the science will be more fully elucidated. In the present state of our knowledge some of these problems are ripe for discussion, others can merely be indicated, while others again have probably remained hidden, though it will be the task of the geologist of the future to clear them up. Among the many questions which demand knowledge of stratigraphical geology for their right understanding are the following, which will be briefly considered in this chapter:—the changes in the position of land and sea in past times, and the growth of continents; the replacement of a school of uniformitarianism by one of evolutionism; and the duration of geological time.

Changes in the position of land and sea. Certain physicists have arrived at the conclusion that the general position of our oceans and continents was determined at a very early period in the earth's history, and that the changes which have occurred in their position since then have been comparatively insignificant. The wide extent of land over which stratified rocks are distributed at once indicates that from the point of view of the geologist the changes have been very important, and it is worth inquiring whether they are not sufficiently important to prove that the primitive oceans and continents have undergone so much alteration as to be unrecognisable. Some authorities, while recognising the great changes which have occurred in the relative position of land and sea during those periods of which geologists have direct information, suppose that the changes took place to a large degree in certain 'critical areas' bordering the more stable areas of permanent ocean on the one side and permanent land on the other.

In discussing the question of general permanence of land and ocean regions it will be convenient to commence with a study of the present land areas, and at the outset we may take into consideration the present distribution of marine sediment over different parts of the land, using the last edition of M. Jules Marcou's geological map of the world for the purpose[1]. A glimpse at this map indicates that more than half of the land areas are occupied by rocks which are as yet unknown (many of which *may* be marine sediments), or by crystalline schists of which the mode of origin has not yet been fully explained, so that a large part of Central Asia, the interior of Africa, and of South America may have existed as land from very early times, and the same may be said of smaller portions of Europe and North America. Actual observation of a geological map therefore indicates the possibility that about half of the land surfaces may have existed as such through very long periods, but though there is a possibility of this, the probability is not very great. The unknown regions, as remarked above, may consist to a

---

[1] A reduced copy of this map will be found opposite the title-page of the first volume of Prof. Prestwich's *Geology*.

considerable extent of marine sediments, and the existence of isolated patches of late Palæozoic and of Mesozoic strata in the heart of Central Asia, points to the sub-mergence of much wider regions than those in which these isolated patches have been found. Again, the character of the sediments when they abut against the crystalline schists frequently proves that these sediments once extended further over the crystalline schists, and have since been removed by denudation, so that even if we assume that the crystalline schists are all of very early date, and not necessarily formed in any case from marine sediments, we cannot suppose that all the area occupied by them has existed as land for long periods of time. On the other hand, the major part of Europe and North Africa, extensive tracts in Asia, the greater part of Australia, a very large part of North America and con-siderable tracts of South America give proofs of having been occupied by the oceans in Palæozoic and later times.

It may be answered that most of these regions con-taining marine sediments occur in critical areas, which have undergone a certain amount of oscillation owing to earth-movements, and that the interior parts of the great continental masses have been practically stationary. But if these lands had been land-areas through geological ages they must have been acted upon by the agents of subaerial denudation, throughout these ages, and long ago reduced to peneplains[1] unless the action of these subaerial agents was counteracted by that of elevating forces; but if these forces were sufficient to counteract the action of subaerial denudation through countless ages

[1] A term proposed by Prof. W. M. Davis for a nearly level surface of subaerial denudation, as opposed to a plain of marine denudation.

they were also sufficient to raise extensive tracts of land above sea-level, and materially to alter the distribution of land and sea, and if elevation could go on to this extent, why not also depression?

Proceeding a step further, and examining the character of the sediments as well as their geographical distribution, we find further evidence of great crust-movements. It has been urged that deep-water sediments do not occur amongst the strata found on the continents,—that there are no representatives of the abysmal deposits of recent ocean floors amongst the strata of the geological column[1], but the researches of the last two decades have brought to light foraminiferal and radiolarian deposits, pteropodal deposits, and possibly deep-sea clays, which are comparable with those in process of formation at great depths in existing oceans, and though the proofs of their deep-sea origin are not always as full as might be desired in the case of the older rocks[2], we can speak with greater certainty when we examine those of Tertiary age, and if the deep-sea accumulations of this late date can be uplifted above sea-level, this is much more likely to have occurred with those of past times.   When a deposit like the radiolarian rock of Barbadoes, the deep-water character of which has been conclusively proved, can be elevated into land since Miocene or possibly Pliocene times, it is evident that the crust-movements have been sufficient to produce the most profound changes in the distribution of land and sea during the long ages which are known to us. Another argument against the occurrence of extensive changes has been derived from an examination of those islands which are spoken of as oceanic islands.   Strictly

[1] See Mr A. R. Wallace's *Island Life.*
[2] See chapter IX.

speaking an oceanic island is one in which the prese
fauna and flora give indications of their introduction I
transport across intervening sea, and no indications of tl
existence of forms of life which inhabited it when it w
once united to a continent; it may be inferred with
considerable degree of certainty that these islands ha
been isolated for long periods of time. It has been stat
that these oceanic islands never contain marine sedimen
of any considerable degree of antiquity, and that there a
therefore no traces of former continents over those wi
tracts of ocean which are occupied by oceanic island
The evidence is of a negative character. The islan
would be less likely to exhibit ancient sediments th
continents, for being near the ocean, they would be readi
submerged, and the older deposits masked by newer on
though this need not necessarily account for the enti
absence of ancient rocks amongst them. The danger
the argument lies in the fact that we do not yet kn
how far these old rocks really are absent, as the geolo
of the oceanic isles has not been fully explored from tl
point of view, and already several cases of the assert
presence of ancient rocks on these islands have be
recorded.

The argument derived from the present distributi
of organisms is far too complex to be discussed here, a
the student is recommended to read a masterly review
the evidence in Dr W. T. Blanford's Presidential Addr
to the Geological Society in 1890, on the question of t
Permanence of Ocean Basins[1]. After reviewing t
evidence furnished by a study of modern distribution
concludes that it " is far too contradictory to be receiv
as proof of the permanence of oceans and continents."

[1] *Quart. Journ. Geol. Soc.*, vol. XLVI., *Proc.*, p. 59.

The existence of former extensive land tracts over regions now occupied by sea is naturally more difficult to prove than that of sea over land, as we depend upon inference rather than actual observation to a much greater degree than when considering the permanence of continents, nevertheless a considerable amount of indirect evidence in favour of the existence of widespread land tracts over our present ocean regions has been accumulated and will be briefly noticed. We may take first the evidence derived from the nature of sediments, and afterwards that which has been acquired by studying distribution of organisms in past times.

The indications of existence of an extensive tract of continent over the North Atlantic Ocean, during Palæozoic times have already been considered, and it was seen that the thinning out of the Palæozoic sediments when traced away from the present Atlantic borders in an easterly direction over Europe and in a westerly one over North America pointed to the existence of this Palæozoic 'Atlantis,' as maintained by Prof. Hull in his work, "Contributions to the Physical History of the British Isles." This writer gives some reasons for supposing that the continental mass began to break up towards the end of Palæozoic times, though it is not clear that complete replacement of land by sea occurred, and the nature of the Wealden deposits has been pointed to as evidence of the existence of an extensive tract of land to the west of Britain during the Cretaceous period.

The Palæontological evidence in favour of destruction of ancient continental areas and their replacement by the sea is more satisfactory than that which is based on physical grounds. The distribution of the Glossopteris flora of the Permo-Carboniferous period points to the former existence of a great southern continent, including the sites of Australia, India, South Africa and South America,—the Gondwanaland of Prof. E. Suess[1].

Again, a study of Jurassic and Cretaceous faunas has led palæontologists to conclude that there was a connexion betwixt S. Africa and India in Mesozoic times across a portion of the area now occupied by the Indian Ocean, and also between S. Africa and S. America, and these inferences are supported by study of the distribution of existing forms.

The sudden appearance of the Dicotyledonous Angio-sperms in Upper Cretaceous rocks has also been used as evidence of destruction of considerable tracts of land subsequently to Upper Cretaceous times, and there is a certain amount of evidence in favour of the existence of this land in the north polar region, in an area now largely occupied by water, though relics of it are left, as the Faroe Isles, Spitsbergen, Novaya Zembla and Franz Josef Land.

I cannot conclude the consideration of the question of permanence of oceans and continents more fitly than by quoting from Dr Blanford's address. He says, "There is no evidence whatever in favour of the extreme view accepted by some physicists and geologists that every ocean-bed now more than 1000 fathoms deep has always been ocean, and that no part of the continental area has ever been beneath the deep sea. Not only is there clear proof that some land-areas lying within continental limits have at a comparatively recent date been submerged over 1000 fathoms, whilst sea-bottoms now over 1000 fathoms deep must have been land in part of the Tertiary era,

[1] On this question and that of the other destroyed continental areas noted here, see W. T. Blanford's *Presidential Address, loc. cit.*

but there are a mass of facts both geological and bio-logical in favour of land-connexion having formerly existed in certain cases across what are now broad and deep ocean[1]."

Growth of continents. Whatever view as to the general permanence of continents and oceans be ultimately established, the occurrence of widespread changes in the position of land and sea is indisputable, and it is of interest for us to consider the nature of these changes in the formation of continents. Prof. J. D. Dana has put forward a hypothesis of growth of continents by a process of accretion, causing diminution in the oceanic areas, which at the same time became deeper: such growth need not always take place in exactly the same way, and study of the distribution of the strata of the North American continent suggests that the growth there was endogenous, the older rocks lying to the west and north forming a horseshoe shaped continent enclosing a gulf-like prolongation of the Atlantic, which became contracted by deposition and uplift in successive geological periods, though it is still partly existent as the Gulf of Mexico. The Eurasian continent, especially its western portion, suggests more irregular growth around scattered nuclei of older rocks, though the process is not completed, and many gulf-like prolongations, as the Baltic and the Mediterranean, still remain as water-tracts, which have not yet been added to the continents.

Although extensive additions to continents may be and no doubt are often largely due to epeirogenic movements, the influence of orogenic movements on continent-formation is very pronounced. As the result of orogenic movements, the rocks of portions of the earth's crust

[1] *Loc. cit., Proc.* p. 107.

become greatly compressed, and give rise to masses which readily resist denudation; moreover, these comparatively rigid masses, as shown by M. Bertrand, tend to undergo elevation along the same lines as those which formed the axes of previous elevations, and accordingly after a continental area has undergone denudation for a considerable period, the uplands consist of rocks which have undergone orogenic disturbance, while the tracts of ground which are occupied by rocks which have not suffered disturbances of this character, even if originally uplifted far above sea-level, tend to be destroyed, and ultimately occupied by tracts of ocean. Stumps of former mountain chains may be again and again established as nuclei of continents and as every period of orogenic movement will add to the number of these nuclei, the continental areas must in course of time become more complex in structure. Moreover, as some areas are affected by orogenic movements to a greater extent than others, the complexity of different continental masses will vary. Thus, western Europe has been affected by orogenic movements during many periods since the commencement of Cambrian times and its structure is extremely complex, while the central and western parts of Russia have not been subjected to violent orogenic disturbances since Cambrian times, and accordingly we find the structure of that area comparatively simple; the greater part of Africa seems to have escaped these movements since remote times, and the structure of that continent is extremely simple when compared with the Eurasian continental tract. It need hardly be stated that the formation of extensive chains composed of volcanic material, by accumulation of lavas and ashes on the earth's surface, may give and often has given rise to more rigid tracts, which will bring about the same effects as those produced by orogenic disturbance as illustrated on a small scale by the Lower Palæozoic volcanic rocks of Cambria and Cumbria.

Uniformitarianism and Evolution. According to the extreme uniformitarian views held by some geologists, the agents which are in operation at the present day are similar in kind and in intensity to those which were at work in past times, though no geologist will be found who is sufficiently bold to assert that this holds true for all periods of the earth's history, but only for those of which the geologist has direct information derived from a study of the rocks, and he is content to follow his master Hutton in ignoring periods of which he cannot find records amongst the rocks. The modern geologist, however, while rightly regarding the rocks as his principal source of information finds that he cannot afford to ignore the evidence furnished by the physicist, chemist, astronomer and biologist, which throws light upon the history of periods far earlier than those of which he has any records preserved amongst the outer portions of the earth itself, just as the modern historian is not content with written records, but must turn to the 'prehistoric' archæologist and geologist for information concerning the history of early man upon the earth. Interpreting the scope of geology in this general way, rigid uniformitarianism must be abandoned. Assuming that the tenets of the evolutionist school are generally true, the question is, how far does this affect the geologist in his study of those periods of which we have definite records amongst the rocks? This is a question which cannot readily be answered at the present day, for our study of the rocks is not sufficiently far advanced to enable us to point out effects amongst the older rocks which were clearly caused by agents working with greater intensity than they do at present, but as, on the other hand, we cannot prove that these effects are due to agents working with no greater intensity than that which now marks these operations, it is unphilosophical to assume the latter. No student of science at the present day would state that because there has been no observed case of incoming of fresh species within the time that man has actually observed the present faunas and floras, the hypothesis of evolution of organisms is disproved, for the time of observation has been too short, and similarly the time which has elapsed since the formation of, say, the Cambrian rocks may have been too short, as compared with the time which has elapsed since the formation of the earth, to allow of any important change in the operation of the geological agents.

Leaving out of account, for the moment, the actual evidence which has been derived from a study of the rocks, we may briefly consider the theoretical grounds upon which the substitution of an evolutionist school of geology for one of uniformity has been suggested[1]. The principal sources of energy which have exerted an influence upon geological changes are the heat received from the sun and that given off from the earth itself, both of which must have diminished in quantity throughout geological ages. To the former source we largely owe climatic changes and the operations of denudation, and accordingly of deposition; to the latter, those of earth-movement and vulcanicity. It by no means follows that because the agents were once potentially more powerful than now, they would necessarily produce greater effects, for that depends to some extent upon the various

[1] The student may consult an interesting article by Prof. Sollas bearing on this subject. See *Geol. Mag.* Dec. 2, vol. IV. p. 1.

conditions which prevailed at different times. To give an example:—if there had at any time been a universal ocean of considerable depth, however active the agents of denudation were then, they could produce no effect whatever, having nothing to work upon; to take a less extreme case, if our continents at any past time were smaller and less elevated than at present, agents of denudation working with greater intensity than that of the present agents need not necessarily have produced a greater amount of denudation than that which is going on at the present day. Again, let us consider vulcanicity: " It is as certain," says Lord Kelvin, " that there is less volcanic energy in the whole earth than there was a thousand years ago, as it is that there is less gunpowder in a ' Monitor' after she has been seen to discharge shot and shell, whether at a nearly equable rate or not, for five hours without receiving fresh supplies than there was at the beginning of the action." But it does not follow that the manifestations of volcanic activity were necessarily more violent in early geological times than now, for the degree of violence would be affected by other things than the volcanic energy, such as the thickness of the earth's crust.

And now, let us consider briefly the characters of the rocks of the crust, to see if they throw any light upon this question. The earliest sediments of which we have any certain knowledge resemble in a striking manner those formed at the present day, and they seem to have been formed under very much the same conditions, though further work may show that there were somewhat different conditions which did produce definite differences in the characters of the earlier strata[1]. Our knowledge of earth-

[1] On this matter see Teall, J. J. H., 'Presidential Address to Section C,' *Report of the British Association*, 1893.

M.

movement and vulcanicity which took place in past ti
is still too small to enable us to draw any certain
clusions connected with the subject under discussion 1
it.   Perhaps the most suggestive indication of one s
conditions having been generally similar in those ε
periods of which we have definite records amongst
rocks is furnished by study of past climate.   If
accept the nebular hypothesis as a starting point,
must admit that in the early stages of the earth's his
the temperature of the surface, which would then
largely dependent upon the amount of heat given
from the earth itself as well as upon that received 1
the sun, must have been much higher than it is at
present day, and indeed the mere diminution of
amount of heat received from the sun would probabl
sufficient to account for a very marked lowering of
temperature.   Besides this change of temperature, resul
in gradual lowering of temperature over the whole ea
surface, we have other changes dependent upon diffe
conditions, as proved by the fact, that there have 1
alternations of glacial and genial periods.   If the gei
temperature had been very high in the early perioc
which we have actual records, the oscillations would
be sufficient to produce a lowering of temperature suffi
to cause glacial periods, whereas if it had not
appreciably higher than now, glacial periods migh
produced.   This may be represented diagrammatica

Let a represent the temperature at the comm
ment of earth-history and b that necessary for glacis
and bc the lapse of time between then and now.
curved line indicates the gradual fall in temper
due to diminution of the amount of heat, while
zigzag line represents the oscillations due to se

climatic changes. If the Cambrian period x occurred comparatively soon after the commencement of earth-history as shown in fig. A, no glaciation could be produced, even during periods when secular changes caused colder conditions than the mean, whereas if the Cambrian period occurred at a time very remote from the commencement of earth-history as shown in B, glacial conditions could be produced then as now, for the mean temperature, as shown by the distance of the curve from the line bc, would be prae-

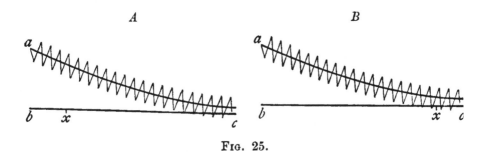

FIG. 25.

tically as it now is. The studies of the last few decades have brought into prominence the occurrence of glacial periods in remote times, probably in early Palæozoic times; and as far as the mean temperature of the earth's surface is concerned, it would appear, from the knowledge in our possession, that matters were not very different in those early times from what they now are.

Some further remarks will be made in subsequent paragraphs concerning the period of the earth's history at which the geologist is first furnished with definite records, but in the meantime it may be observed that the geologist will do well, when working amongst the strata, to consider that the more active operation of agents, even in times of which he has definite knowledge, may have produced effects which he should be prepared to discover,

as their discovery would be of considerable import&
and that he should not be content to infer that bec
it has been proved that agents operating with the &
intensity as that which they have at present, *may*
produced all the effects which he can actually obs&
they therefore necessarily *did* produce them.

Recurrences.   Absolute uniformity of conditior
impossible, even in a single area.   Every change w
takes place upon the earth produces conditions some·
dissimilar from those which previously existed, and t
will leave their effects upon the physiography of the ﹔
For this reason, assuming that the conditions have gr
ally changed from simpler to more complex, every p&
of time will have been marked by conditions which r
prevailed before or afterwards, and these will leave ·
impress upon the deposits of the period.   It is dou
for instance, as already remarked, whether the &
conditions which gave rise to the extensive deposi
vegetable matter in Carboniferous times which now
coal, ever occurred to a like extent in previous or
sequent periods, and accordingly, though we have def
of coal of other ages, none are so extensive as those c
Coal Measures.   Again, as the strata of one perioc
largely composed of denuded particles of pre-exi
strata, which were derived directly or indirectly
igneous rock, the soluble material existing in the ig:
rocks must have been gradually eliminated unless res
by other processes, and we might expect to find
early sediments have, on the whole, a larger proport
soluble silicates than the later ones.

Besides these changes, there are physical ch
which are recurrent, and cause conditions generally s:
to pre-existing ones to occur in an area after an inter

dissimilar ones.   We have seen that deposits tend to vary according to the distance from the coast, limestone being succeeded by mud, this by sand and gravel, and after subsidence the sand and gravel are succeeded by mud, and that by limestone. These changes will produce some effect upon the organisms, and the recurrence of organ-isms is a well-known event, of which cases have been cited in a former chapter.

Again we find, as already pointed out, recurrence of climatic changes, with alternation of glacial and warmer periods, and these may have been very widespread, and would influence the other physical conditions, as well as the distribution of the organisms.   Vulcanicity may have been more rife at some periods than others, for instance there seems, in the present imperfect state of our know-ledge, evidence of enfeebled vulcanicity in later Mesozoic times, and of its renewed activity in Tertiary times. Again, orogenic movements seem to have occurred more extensively at some times than others, as for instance in early upper Palæozoic times, at the end of the Palæozoic epoch, and in early Tertiary times, though this may also be an apparent and not an actual truth, due to imperfect knowledge.   In any case, in limited areas, there seem to have been alternations of periods of uplift accompanied by marked orogenic movements, and of widespread depres-sion, accompanied by sedimentation.

The subject of rhythmic recurrence is worthy of further study.   This recurrence in combination with evolutionary change may account for the apparent marked difference between Cambrian and Precambrian times, a difference which strikes some geologists as being too great to be accounted for as due to our ignorance only.

Organic evolution. This subject is too wide for more

than passing notice in a work of this character. The
evidence of Palæontology is of extreme importance to the
biologist, and indeed, the way in which evolution o
organisms has occurred can only be actually demonstrated
by reference to Palæontology, and the study of Palæonto
logy has already given much information concerning the
lines on which evolution has proceeded in different group;
of organisms. It must be remembered that the majo.
divisions of the invertebrata were in existence in very
early times; indeed representatives of most of them are
found in the rocks containing the earliest known fauna
that of the *Olenellus* beds of Cambrian age. If ou
present views as to evolution be correct, there is no doub
that the period which elapsed between the appearance o
life upon the globe and the existence of the *Olenellu.*
fauna must have been very great, possibly, as Huxley
suggested, much greater than that which has elapsed
between early Cambrian times and the present day. I
this be so, however probable it is that we shall carry ou
knowledge of ancient faunas far back beyond Cambrian
times, it is extremely improbable that we shall ever ge
traces of the very earliest faunas which occupied our
earth.

Geological time. Various attempts have been made
to give numerical estimates of the lapse of time which
occurred since the earth was formed, or since the earlies
known rocks were deposited. These attempts may b
classed under two heads, namely, those made by physicists
mainly on evidence obtained otherwise than by a study c
the rocks, and those made by geologists by calculating
the mean rate of denudation and deposition of the rock
and estimating the average thickness of the rocks of th
geological column.

The calculations of physicists as to the age of the earth vary :—Lord Kelvin assigned 20,000,000 years as the minimum and 100,000,000 as the maximum duration of geological time. Prof. Tait has halved Lord Kelvin's minimum period, while Prof. G. Darwin admits the possibility of the lapse of 500,000,000 years.

The estimates made by geologists, which will appeal more directly to the geological student, also vary considerably, though they bear some proportion to those which have been put forward by the physicists. Prof. S. Haughton[1] assigned a period of 200,000,000 years for the accumulation of the rocks of the geological column; Mr Clifton Ward[2] one of 62,000,000 years, after studying the rocks of the English Lake District, and allowing for the gaps in the succession; Mr A. R. Wallace[3] further lowers the time for the formation of the column to 28,000,000 years; Sir A. Geikie[4] gives 73,000,000 years as the minimum and 680,000,000 as the maximum; while Mr J. G. Goodchild has lately[5] estimated the period at over 700,000,000 years.

Interesting as these figures are, they probably convey little to the ordinary reader, and it is doubtful whether the geologist is really affected by them to any extent when picturing to himself the vast duration of geological time. One numerical estimate probably does impress him, namely that made by Croll as to the date of the Great Ice Age, for if the Ice Age be so remote as

[1] *Nature*, vol. XVIII. p. 268.

[2] Ward, J. C., 'The Physical History of the English Lake District,' *Geol. Mag.* dec. 2, vol. VI. p. 110.

[3] Wallace, A. R., *Island Life*, Chap. x.

[4] Geikie, Sir A., 'Presidential Address to the British Association,' *Report Brit. Assoc.*, 1892.

[5] Goodchild, J. G., *Proc. Roy. Soc. Edinburgh*, vol. XIII. p. 259.

Croll imagined, the commencement of earth-history must be inconceivably more remote; as Croll's estimate is not generally accepted, it is doubtful how far geologists are thus influenced, and probably the fact which does impress them most, leaving fossils out of account, is the very little change which has occurred in historic or even in prehistoric times as compared with the vast changes which are familiar to them after studying the strata of the geological column.

It is, after all, the succession of varied faunas which really gives students of the rocks the most convincing proof of the vast periods of geological time. If anyone doubts this assertion, let him consider what impression would be made upon him by observing the several thousand feet of strata of the column if none of them contained any organisms. Cognisant as he is of the slow rate of change of existing organisms, the fact that fauna has succeeded fauna in past times brings home to him in an unmistakeable manner the great antiquity of the earliest fossiliferous rocks, and as our detailed knowledge of these faunas increases the impression of great lapse of time is intensified. And if the earliest fossiliferous rocks be of such vast antiquity, and, as has been remarked, the period of their formation is comparatively recent with reference to the actual commencement of earth-history, the latter must indeed be inconceivably remote, and numerical estimates can do but little to familiarise us with the significance of the vast time which has rolled by since the world's birthday.

INDEX.

CAMBRIDGE: PRINTED BY J. AND C. F. CLAY, AT THE UNIVERSITY PRESS

BIOLOGICAL SERIES.

GENERAL EDITOR, A. E. SHIPLEY, M.A.

Elementary Palæontology—Invertebrate	H. Woods, M.A., F.G.S.	
Elements of Botany	F. Darwin, M.A., F.R.S.	4s.
Practical Physiology of Plants	F. Darwin, & E. H. Acton, M.A.	4s
Practical Morbid Anatomy	H. D. Rolleston, M.D., F.R.C.P. & A. A. Kanthack, M.D., M.R.C.P.	
Zoogeography	F. E. Beddard, M.A., F.R.S.	
Flowering-Plants and Ferns	J. C. Willis, M.A. In two vols.	10s
The Vertebrate Skeleton	S. H. Reynolds, M.A.	12s.
Fossil Plants	A. C. Seward, M.A., F.G.S. 2 ᵥ Vol. I.	
Palæontology—Vertebrate	A. S. Woodward.	

PHYSICAL SERIES.

GENERAL EDITOR, R. T. GLAZEBROOK, M.A., F.R.S.

Heat and Light	R. T. Glazebrook, M.A., F.R.S.	
,, ,, in two separate parts	,, ,,	each
Mechanics and Hydrostatics	,,	8s.
,, ,, in three separate parts		
Part I. Dynamics	,,	
,, II. Statics	,,	,,
,, III. Hydrostatics	,,	,,
Solution and Electrolysis	W. C. D. Whetham, M.A.	7s.
Electricity and Magnetism	R. T. Glazebrook, M.A., F.R.S. [In Preparati	
Sound	J. W. Capstick, M.A. [In Preparati	

GEOLOGICAL SERIES.

Petrology for Students	A. Harker, M.A., F.G.S.	7s.
Handbook to the Geology of Cambs. ...	F. R. C. Reed, M.A.	7s.
The Principles of Stratigraphical Geology	J. E. Marr, M.A.	
Crystallography	Prof. Lewis, M.A. [In Preparati	

Laboratory Note-Books of)
Elementary Physics) ... ... { L. R. Wilberforce, M.A., a
{ T. C. Fitzpatrick, M.A.

I. Mechanics and Hydrostatics.   II. Heat and Optics.
III. Electricity and Magnetism.   .   each

Other volumes are in preparation and will be announced shortl

BIOLOGICAL SERIES.

A Manual and Dictionary of the Flowering Plants a
Ferns. Morphology, Natural History and Classificati
Alphabetically arranged. By J. C. WILLIS, M.A., Gonvi
and Caius College. In Two Volumes. Crown 8vo. W
Illustrations. 10s. 6d.

Bookman. One of the most useful books existing for students of bota
...The student who has this book and the chances which Kew, or even on
the smaller gardens, affords him, will make a steady and really scient
progress.

Gardeners' Chronicle. Altogether the amount of information crow
into these 600 or more small pages is quite wonderful, and attests not o
the assiduity but the accuracy of the compiler....For advanced students
experts these little volumes will be most useful for reference, as accuracy
statement and general freedom from accidental errors are as remarkabl
them as are the comprehensiveness and condensation of treatment.

Elements of Botany. By F. DARWIN, M.A., F.R.S. Seco
Edition. Crown 8vo. With numerous Illustrations. 4s.

Journal of Education. We are very pleased with this little book, wh
teaches without being didactic; and leads us into the various phenomena
plant life with an easy gait, quite free from the halting jolt of the aver
text-book on vegetable morphology and physiology. The style is poin
indeed almost abrupt; but still very clear and concise. The description
the various examples of physiological functions amongst plants is distin
good, and Mr Darwin's method of teaching the most fundamental functi
of plant life from simple experiments with easily get-at-able specimen
thoroughly scientific. It is a noteworthy addition to our botanical literat

Practical Physiology of Plants. By F. DARWIN, M.
F.R.S., Fellow of Christ's College, Cambridge, and Rea
in Botany in the University, and E. H. ACTON, M.A., l
Fellow and Lecturer of St John's College, Cambridge. W
Illustrations. Second Edition. Crown 8vo. 4s. 6d.

Nature. A volume of this kind was very much needed, and it i
matter for congratulation that the work has fallen into the most compe
hands. There was nothing of the kind in English before, and the book
be of the greatest service to both teachers and students....... The thorou
practical character of Messrs Darwin and Acton's book seems to us a g
merit; every word in it is of direct use to the experimental worker an
him alone.......The authors are much to be congratulated on their w
which fills a serious gap in the botanical literature of this country.

Zoogeography. By F. E. BEDDARD, M.A., F.R.S. W
Maps. 6s.

Daily Chronicle. Although included in the series of Cambridge Nat
Science Manuals, and therefore designed chiefly for students of biol
Mr Beddard deals with his subject in a clear and graphic way that sh
commend his book to the general reader interested in the question.
style, while never lacking dignity, avoids the dulness which too
accompanies that virtue.

BIOLOGICAL SERIES.

Elementary Palæontology—Invertebrate. By HENRY WOODS, M.A., F.G.S. With Illustrations. Crown 8vo. Second Edition. 6s.

Nature. As an introduction to the study of palæontology Mr Woods's book is worthy of high praise.

The Vertebrate Skeleton. By S. H. REYNOLDS, M.A., Trinity College. Crown 8vo. 12s. 6d.

British Medical Journal. A volume which will certainly take its place amongst the standard text-books of the day.

Lancet. This work should prove valuable alike to the beginner and to the more advanced student.......We feel sure that the book is well worth the attention of all who are interested in the subject.

Practical Morbid Anatomy. By H. D. ROLLESTON, M.D., F.R.C.P., Fellow of St John's College, Cambridge, Assistant Physician and Lecturer on Pathology, St George's Hospital, London, and A. A. KANTHACK, M.D., M.R.C.P., Lecturer on Pathology, St Bartholomew's Hospital, London. Crown 8vo. 6s.

British Medical Journal. This manual can in every sense be most highly recommended, and it should supply what has hitherto been a real want.

PHYSICAL SERIES.

Mechanics and Hydrostatics. An Elementary Text-book, Theoretical and Practical, for Colleges and Schools. By R. T. GLAZEBROOK, M.A., F.R.S., Fellow of Trinity College, Cambridge, Principal of University College, Liverpool. With Illustrations. Crown 8vo. 8s. 6d.

Also in separate parts.

Part I. **Dynamics.** 4s.   Part II. **Statics.** 3s.

Part III. **Hydrostatics.** 3s.

Educational Times. Mr Glazebrook's excellent elementary text-book combines the theoretical and the practical more successfully than any school-book which we have seen. Its statements are at once plain and sound and many of his directions for experimentation are remarkably apt and striking. The Series of the 'Cambridge Natural Science Manuals' promises to make a new departure in elementary science manuals for school and college.

Educational News (Edinburgh). We recommend the book to the attention of all students and teachers of this branch of physical science.

Knowledge. We cordially recommend Mr Glazebrook's volumes to the notice of teachers.

Practical Teacher. We heartily recommend these books to the notice of all science teachers, and especially to the masters of Organised Science Schools, which will soon have to face the question of simple practical work in physics, for which these books will constitute an admirable introduction if not a complete *vade mecum.*

Press Opinions.

Heat and Light. An Elementary Text-book, Theoretical and Practical, for Colleges and Schools. By R. T. GLAZEBROOK, M.A., F.R.S., Fellow of Trinity College, Cambridge, Principal of University College, Liverpool. Crown 8vo. 5s. The two parts are also published separately.

<table>
<tr><td>**Heat.** 3s.</td><td>**Light.** 3s.</td></tr>
</table>

Nature. Teachers who require a book on Light, suitable for the Class-room and Laboratory, would do well to adopt Mr Glazebrook's work.

Journal of Education. We have no hesitation in recommending this book to the notice of teachers.

Practical Photographer. Mr Glazebrook's text-book on "Light" cannot be too highly recommended.

Solution and Electrolysis. By W. C. D. WHETHAM, M.A., Fellow of Trinity College. Crown 8vo. 7s. 6d.

Electrician. This book is a valuable addition to the literature of the subject.

GEOLOGICAL SERIES.

Handbook to the Geology of Cambridgeshire. For the use of Students. By F. R. COWPER REED, M.A., F.G.S., Assistant to the Woodwardian Professor of Geology. With Illustrations. Crown 8vo. 7s. 6d.

Nature. The geology of Cambridgeshire possesses a special interest for many students....There is much in Cambridgeshire geology to arouse interest when once an enthusiasm for the science has been kindled, and there was need of a concise hand-book which should clearly describe and explain the leading facts that have been made known....The present work is a model of what a county geology should be.

Petrology for Students. An Introduction to the Study of Rocks under the Microscope. By A. HARKER, M.A., F.G.S., Fellow of St John's College, and Demonstrator in Geology (Petrology) in the University of Cambridge. Crown 8vo. Second Edition, Revised. 7s. 6d.

Nature. No better introduction to the study of petrology could be desired than is afforded by Mr Harker's volume.

London: C. J. CLAY AND SONS,
CAMBRIDGE UNIVERSITY PRESS WAREHOUSE,
AVE MARIA LANE.
AND
H. K. LEWIS, 136, GOWER STREET, W.C.
Medical Publisher and Bookseller.

Author _____ Marr, J. E.

Title _____ Geology.